GENOMIC CITIZENSHIP

GENOMIC CITIZENSHIP

The Molecularization of Identity in the Contemporary Middle East

IAN MCGONIGLE

The MIT Press
Cambridge, Massachusetts
London, England

The open access edition of this book was made possible by generous funding from Arcadia—a charitable fund of Lisbet Rausing and Peter Baldwin.

The MIT Press would like to thank the anonymous peer reviewers who provided comments on drafts of this book. The generous work of academic experts is essential for establishing the authority and quality of our publications. We acknowledge with gratitude the contributions of these otherwise uncredited readers.

This book was set in Garamond by Westchester Publishing Services, Danbury, CT.

Library of Congress Cataloging-in-Publication Data

Names: McGonigle, Ian, author.
Title: Genomic citizenship : the molecularization of identity in the contemporary Middle East / Ian McGonigle.
Description: Cambridge, Massachusetts : The MIT Press, 2021. | Includes bibliographical references and index.
Identifiers: LCCN 2020047210 | ISBN 9780262542944 (paperback)
Subjects: LCSH: Human genetics--Political aspects--Israel. | Human genetics--Political aspects--Qatar. | Biobanks--Political aspects--Israel. | Biobanks--Political aspects--Qatar. | Ethnology--Israel. | Ethnology--Qatar. | Ethnicity--Israel. | Ethnicity--Qatar. | Nationalism--Israel. | Nationalism--Qatar. | Citizenship--Israel. | Citizenship--Qatar.
Classification: LCC QH428.2.I75 M38 2021 | DDC 599.93/5--dc23
LC record available at https://lccn.loc.gov/2020047210

CONTENTS

AUTHOR'S NOTE

The text of this book was written solely by the author, except where otherwise stated. Chapter 2 contains elements of a paper written in collaboration with Lauren Herman, a shorter version of which was published in the *Journal of Law and the Biosciences* (McGonigle and Herman 2015). Chapter 2 also includes parts previously published in *Transversal: Journal for Jewish Studies* (McGonigle 2015). A comparative article drawing on material from chapters 3 and 4 was published by *Science, Technology and Society* (McGonigle 2020b), and some of the material in chapter 4 is included in a chapter in an edited volume on identity and authenticity published by Routledge (McGonigle 2020a). Any previously published material is reproduced here with permission from the publisher. Likewise, permission has been granted for the use of any copyrighted images or figures.

LIST OF TABLES AND FIGURES

PREFACE

In 2019, then–Lebanese foreign minister Gebran Bassil tweeted about the genetic uniqueness of the Lebanese: "We have devoted a concept to our Lebanese identity, above any other affiliation, and we have said that it was genetic, since it was the only explanation for our similarity and distinction" (Nassar 2019). When challenged by British journalist Stephen Sackur on BBC *HARDtalk* (2019) for suggesting Lebanese genetic superiority to the Syrians, Bassil dismissed the accusation of racism but doubled down on the concept of "Lebanity," which, he said, offers "a sense of belonging for the Lebanese," adding, "We have our own belonging, not being affiliated to any other foreign country."

Genetics has crept into how we think about national uniqueness and ethnic exceptionalism. How has this happened? *Genomic Citizenship* seeks to explain how biology—and specifically genetics—has become part of ethnonationalism's conceptual walls of national inclusion and exclusion. More precisely, *Genomic Citizenship* seeks to explain how and why genetics is growing as a genre for imagining ethnonational belonging in the Middle East.

Genomics research enabled by the collection of biological samples from ethnic populations is an important part of global medical research aimed at improving health and eliminating genetic disease, but the resultant technologies also provide a way of imagining and performing national citizenship. The consequences for how ethnic and national identities are understood are significant, with each national context having unique aspects. Comparing the genetics and biobanking of ethnic populations in Israel and

Qatar underpins the core thesis of this book: that the molecular realm is an emergent site for articulations of ethnonational identity in the contemporary Middle East.

Increasingly, the nation itself is imagined as a biological object. When people imagine their relationship with their fellow citizens, they draw now on the language of genetics. "It's in our DNA" has become a catchphrase used in a way that can be half-serious or metaphorical, but it is also indicative of widespread acceptance of genetics as a basis of identity. By comparing how inclusion in the nation is imagined and performed in the Middle Eastern ethnonations of Israel and Qatar, this book attempts to clarify the theoretical problematic of the relationship between the science of ethnic populations and the valuating national context of its emergence. The implicit political normativity of this line of thinking is a commitment to keeping possibilities open and to unsettling reified ethnonational identities and taken-for-granted identitarian imaginaries. The intellectual value of this critical approach lies in the power of possibility that a responsible criticism of the relationships between science and society renders. In this sense, this reading of science and society entails an implicit proposition of "the otherwise," wedging open a gap of possibility between science, as the self-imagined reportage of "what is," and politics, understood as the battleground of "what could be." An effective critique of science ought to shift the balance toward the latter.

Since this is a book about the relationship between science and identity, as well as a book that draws in part on the author's personal interactions with scientists, you are likely to wonder about my own identity and my relationship to science. I became interested in philosophical and anthropological questions about science while I was working toward a PhD at one of the world's most prestigious biochemistry departments, at the University of Cambridge. I was engaged full time in laboratory research on the structure and function of brain neurotransmitter receptors. By day I was on a trajectory to become a career research scientist, while in my free time I read authors like Michel Foucault, Pierre Bourdieu, Herbert Marcuse, Mary Douglas, Claude Lévi-Strauss, Philippe Descola, classic British structuralist and functionalist ethnographies, and accounts of Amazonian shamanism. I became preoccupied by questions of epistemology, ontology, and how scientific objects—and comparatively, diffuse "indigenous" objects, like plant spirits—can appear and disappear, in different historical periods, or indeed simultaneously, synchronically, cross-culturally. In short, I came to the question of what the cultural basis is of a scientific object.

Around this point, my fascination with philosophical anthropology melded with a curiosity about the Middle East, and I began to take decisive strides to train as an anthropologist and pursue questions about nationalism, identity, and epistemology in the Middle East.

In bridging the gulf between my training as a bioscientist and my nascent fascination with anthropology, I began thinking about what a reflexive philosophical anthropology of scientific objects would look like. Specifically, I started seriously considering the practice of cultural translation in the field of ethnopharmacology, the study of indigenous drugs. This topic became the basis for my master's thesis at the University of Chicago, which was later published as an essay in *Ethnos* (McGonigle 2017). While working on the topic, I was pursuing a theoretical understanding I had been grappling with since the time I had worked in the laboratory. This involved a proposition of epistemological "symmetry" and a commitment to a type of responsible critique that contextualizes science with politics, ethics, morality, and other human values. Rather than pitting indigenous facts against scientific facts, in an asymmetric global power relation that would necessarily impugn nonscientific ways of knowing, I became interested in the relationships between facts and the politics of their respective contexts. In essence, this brought me to an anthropological reading of science as a political discourse of world building, and I found my intellectual niche at the productive intersection of philosophical anthropology and science, technology, and society (STS).

While taking a class on sacred spaces in contemporary Israel with visiting Professor Yoram Bilu (of the Hebrew University) at the University of Chicago, I became increasingly interested in Israel and the Middle East. I decided shortly afterward that the ethnic-rich Middle East would be a fitting site for a contemporary study of science and society. I wrote a research proposal that focused on the role of ethnic genetics in Israel and, in preparation for fieldwork, I began learning Modern Hebrew. While spending a year as a visiting fellow at Harvard's STS program, where I came to hone my STS thinking in relation to my doctoral project, I formed a working relationship with three exceptional professors, an opportunity that motivated me to continue my studies at Harvard.

After securing a generous postdoctoral fellowship funded by the Israel Institute to do fieldwork on ethnic genetics in Israel and its broader implications for Israeli society, I moved to Tel Aviv for a year of fieldwork. During this time it became clear that extending the question of ethnic genetics to a second field site would sharpen the project and bring into focus the way in which the political context has specific manifestations in the science and its appropriation by broader society. I surveyed other small ethnic states in the Middle East that have significant biomedical research developments, and I decided to extend my project to include a comparison with biobanking developments in Qatar, where the ethnic Qataris constitute a demographic minority, and where the state is heavily investing in the biosciences. I made a research trip to Doha to collect data for this portion of the project, and after discovering a national biobank and national genome project in Doha, the project became a comparative study of Israel and Qatar.

The advantage of comparing Israel and Qatar inheres in the peculiar balance of their similarities and differences. They are two of the most enigmatic nation-states in the region, both regionally isolated and with poor diplomatic ties to their neighbors. This factor alone drives a sense of national insidership. And while Israel is a Jewish-majority state with an explicit Jewish character, Qatar's population is mostly composed of noncitizens with Qatari nationals in the minority. This situation raises the question as to what ethnonationalism looks like in these reciprocal demographic arrangements. On the other hand, Qatar and Israel share the common history of being post-British protectorates with relatively newly established but strong national identities and multiracial populations. Moreover, both states are highly advanced in the biosciences compared to other small Middle Eastern states. These factors combine to make Israel and Qatar ideal candidates for a comparative case study of the mutual constitutions of science and the Middle Eastern ethnonation in the era of genomic citizenship.

This book is more an anthropology of scientific objects than an anthropology of scientists or an ethnography of the laboratory. It centrally concerns the way in which elusive metaphysical imaginaries like "the nation" or "ethnicity" appear or disappear in the epistemic products and consequences of scientific activity. Crucially, I focus on the constitutive societal relationships that frame this process, and the attention here is on the imaginations of peoplehood as they are woven in and out of scientific practice.

The project is, however, built on traditional anthropological methods. Participant observation has been used as an ethnographic method by anthropologists since at least the early twentieth century and has been sustained as a central method of inquiry in the discipline since then (Helmreich 2007). More relevantly, participant observation has been used to do "symmetrical anthropology" by studying scientists ("studying up" is a turn of phrase often used, as opposed to studying native, traditional, societies) in laboratories (Latour and Woolgar 1986). Its strength lies in producing a qualitative understanding of the site studied, allowing the researcher to gain an intimate understanding of the culture from within. I was well prepared to do this particular line of research, having had training in biochemistry, cell biology, biophysics, and neuroscience. That training facilitated smooth integration with scientists as a collaborator and, therefore, also, as a participant observer in the laboratories at Tel Aviv University and at scientific institutions and conferences in Doha. I already knew how to speak like a native, the dominant field language being the language of science.

This book emerges from a one-year participant-based ethnography of the National Laboratory for the Genetics of Israeli Populations combined with three research trips to Qatar totaling about six weeks. In Israel, I worked in a genetics laboratory in the Sackler School of Medicine at Tel Aviv University as a scientist and ethnographer, attending to how genetic readings relate to the cataloging of ethnic identities. In Qatar, I had several conversations with leading researchers and clinicians working in genomic medicine in the Gulf region, attended three functional genomics conferences, conducted targeted interviews, visited key institutions, and conversed with clinicians and scientists in Doha. This research also draws heavily on published documents and reports from the National Laboratory for the Genetics of Israeli Populations, from Sidra Medicine in Doha, from Qatar Biobank, and from the Qatar Genome Programme.

As a trained scientist doing research among scientists, in laboratories, and at scientific conferences, fieldwork often felt like "anthropology at home." As a scientist, I was a "relative native" working with scientists and studying science from an anthropological perspective. Anthropologist of Amazonia Eduardo Viveiros de Castro (2013, 473) describes the relative stance of the anthropologist in the field: "The 'anthropologist' is a person whose discourse concerns the discourse of a 'native.' The native need not be overly savage, traditionalist nor, indeed, native to the place where the anthropologist finds him. The anthropologist, on his part, need not be excessively civilized, modernist, or even foreign to the people his discourse concerns."

What is important is that there is a relationship between the anthropologist and his informants that generates knowledge. I embodied both a partial native and a visitor role in the field. Indeed, the anthropologist need not be entirely native, or foreign. What matters is that the anthropologist's partial identity generates a relation of knowledge production. As a member of the tribe of scientists, I entered the field self-presenting as a scientist-anthropologist, with an acquired identity that was hybrid, partial, and indeed "relatively" native. But in Israel, I was not Jewish, and in Qatar, I was not Muslim. As an Irish Catholic, I was foreign enough to be a complete outsider, but as a trained scientist with a PhD in biochemistry I was a credible interlocutor for the scientists I met. This partial insidership and outsidership has had an impact on the nature of my fieldwork and an especially significant methodological one in terms of my participation. It influenced the nature of my relationship with my informants, a heavy ethical influence in terms of what I brought and asked for. It also shaped the quality of the data collected, a definitive empirical effect on how data are acquired. These issues will become evident in the course of this book.

This project also forced me to engage in a sustained consideration of what it means to be a "good" anthropologist of science. How closely must or can one align (politically, professionally, personally) with the people one studies? To what extent does it help to be in the lab with scientists, as opposed to the interviewing-and-leaving approach? I have worked with my informants at length, and the relationship has been somewhat symmetrical and reciprocal: I have invited them to academic conferences, and I have coauthored scientific papers with them. Moreover, I have taken seriously the responsibility of maintaining ethical relationships with informants, recognizing and sometimes aligning with their interests. Working with scientists in both Tel Aviv and Doha has been a two-way relationship, and I hope to maintain collaborative links in the future and to develop shared projects. For example, I have had discussions with scholars in Doha about working with them on their bioethics protocols. Moreover, when I co-organized a two-day symposium on "the molecularization of identity" at Harvard in April 2016, I invited two of the scientists I worked with in Tel Aviv, David Gurwitz and Noam Shomron, to present their views on biobanking and genetic privacy. Since becoming involved with a human genome project in Singapore, named GenomeAsia100K, I have again been in conversation with scientists in Tel Aviv and Doha about possible collaborations. I have found it to be both personally rewarding and methodologically insightful to engage with genomic scientists both as actors in my domain of study and as teachers and experts in their fields. In this regard, a certain practical "symmetry" has been achieved wherein my anthropological research has become dialogical. I take pride in the fact that my informants are very much active voices in the conversation, and I am humbled by the expertise that speaks back to me when their generous knowledge duly corrects my assumptions.

A word on disciplinarity or, rather, a caveat on the genre. This book engages anthropological theory, STS, intellectual history, critical theory, Middle Eastern studies, cultural studies, and critical legal studies. At the same time, this is not a traditional ethnography of the laboratory. Although I spent many months in the lab in Tel Aviv gaining essential data, and through these inroads experienced countless rich ethnographic moments, some of the most persuasive data that inform this book have come from public records, legal and historical sources, published scientific papers, institutional reports, websites, and brochures. I make my argument with the purposeful juxtaposition of material from diverse sources: image, text, ethnographic moments, documents, downstream research outputs, and the formal legal discourse of the state, with an attentiveness to the genealogy of ideas and their dependence on technologies. This "wide horizon" approach is necessary for the apprehension of scientific objects. Scientific objects ride various channels of mediation to stabilize their ontologies in the present (texts, institutions, laws, graphics, and so on). This form of writing is a likely consequence of pursuing a diffuse anthropological topic such as "ethnic genetics," which is not definitively grounded in a single institution, locale, or set of actors.

The form of this text is also commensurate with the interdisciplinary training I have received along the way: I have been an affiliate in the Harvard STS Program for several years, where I have participated in seminars and workshops dealing with science and society in an interdisciplinary context. Moreover, my time at the Edmond J. Safra Center for Ethics at Tel Aviv University consolidated a commitment to thinking "around" societal *problematiques* rather than advocating a normative position from "within." My interest and dedication to the anthropology of science were nurtured on continental critical theory and philosophical anthropology, and these diverse influences shape the style of reasoning arrived at here. Ultimately, though, I am studying science as culture, and attempting to ground theoretical issues in concrete and localized ethnographic spaces.

In studying science as culture, I have been guided especially by the work of John and Jean Comaroff. In *Ethnography and the Historical Imagination*, they contend that "ethnography serves at once to make the familiar strange and the strange familiar, all the better to understand them both" (1992, 6). For an ethnography of science, this process of "making strange" demands not engaging as a full native in the laboratory, but partially remaining in a zone of ambivalence, focusing on the historical determinants that have rendered the present conditions normative, and not taking the status quo for granted. Concerning the naturalized cosmology of science, Comaroff and Comaroff elaborate: "It is arguable that many of the concepts on which we rely to describe modern life—statistical models, rational choice and game theory, even logocentric event histories, case studies, and biographical narratives—are . . . our own rationalizing cosmology posing as science, our culture parading as historical causality" (6).

For this reason, they argue for what they call a "genuinely historicized anthropology" (Comaroff and Comaroff 1992, 6). This historicized anthropology means digging into the specificities of how the present was arrived at, and interrogating the political a priori that renders the present possible. Extending these insights to an ethnography of epistemic practices themselves—in this case, the biosciences—demands the uncoupling of fact from value. Ethnography itself, of course, does not achieve epistemic supremacy, a God's-eye view. Rather, ethnography itself is replete with epistemological uncertainty, but this condition of ethnography, Comaroff and Comaroff argue, "personifies, in its methods and models, the inescapable dialectic of fact and value" (1992, 9). An ethnography of science must, therefore, involve unsettling the taken for granted; it ought to be a true historicization of facts, a telling of the story of the relation of scientific fact to its own genealogy.

These insights, this research, and my ethnographic experience tracking the phenomenon of the "molecularization of ethnicity" raise an open question about the method of the ethnography of scientific objects. The question, beyond the scope of this book, pertains to the location of the scientific object, in this case, ethnic genetics. In one sense, I am studying the scientific object as a window into the historical construction of the societies I am studying. The rationalizing cosmology of nationalism, it is presupposed, can be apprehended through the study of molecular genetics. Such a theoretically ambitious proposition raises methodological questions: How does one study something as diffuse and multi-sited as "ethnicity" in relation to science, and indeed, reciprocally, "science" in relation to ethnicity? The issue is one of scale, location, and modality of attentiveness. The question also leads to a dialectic, or a "co-production," of science and society (Jasanoff 2004). At issue is the way one finds the broader context in the minutiae of scientific discourse, and likewise how one tracks the wider social life of scientific practice outside the laboratory. I offer this question at the outset to problematize the notion of ethnographic location and to gesture toward a conversation about the productive intersection of anthropology and STS.

But science is a globalized discourse, and one must oscillate between the general and the particular, the global and the local when reading science qua culture. Attention to scale must be varied and wide, simultaneously local and global. Moreover, scientific objects do not speak for themselves, but are of course constituted through human mediations and in diverse ontological registers. In this work, I am attempting to put into focus the broad social implications of science and technology. I am framing the role of science in society at large, and consequently, the focus must extend more broadly than the laboratory itself. This dialectic of context and content demands that one capture the ways in which science is appropriated within society, and likewise one must tackle the ways societal particularities, like ethnic imaginaries, bleed into scientific discourse. For these concrete reasons, this text departs from the conventions of ethnography. I show how the ethnonation appears in scientific discourse in various ways and at varied sites, be they texts, policies, reports, visual culture, or controversies. In this regard, the approach remains somewhat "experimental," to borrow an indigenous concept from the field.

This book is the product of ethnographic research conducted in Israel and Qatar between July 2015 and December 2018. Chapter 1 introduces the conditions of possibility in the biosciences and in global precision medicine developments that have laid the ground for biological measurements of ethnic identity to take root.

In chapter 2, I discuss the origins of biological understandings of Jewishness, especially in relation to the intellectual history of Zionism and Jewish political thought, and the possibility of a novel application of genetics in assigning Israeli citizenship. This chapter, written in the genre of historical anthropology, situates "Jewish genetics" within the diverse political philosophy of Zionism, particularly as it relates to configurations of Jewish ethnicity and modes of imagining citizenship. It discusses this potential biopolitical regulatory technique in the Israeli context, and highlights the implications for citizenship law and defining the limits of belonging in the Jewish nation.

In chapter 3, I present my findings from my ethnographic work at the Israeli biobank, the National Laboratory for the Genetics of Israeli Populations, and I examine the origins, motivations, and aspirations of the Israeli biobank, asking what kind of moral community the biobank mediates. I show that despite my expectations of finding nationalist imaginaries at the level of the laboratory, the Israeli biobank does not explicitly foster or emphasize exclusive Jewish peoplehood but more closely hews to global biobanking trends, which are now becoming dominated by genomic databasing and big data projects. I find that population genomics of Jewish groups in Israel is now being transformed by emerging trends in personalized medicine with the emergence of genetic data as a site of reified economic value.

In chapter 4, I discuss Qatar Biobank and developments in genetic medicine in Qatar, and I analyze the relationship between the Qatari biobank and the context of an emerging Qatari nationalism and state-building projects. I ask what kinds of collective identity and moral community are imagined and fostered through these projects. I argue that the particular tribal history of Qatar is giving way to a new national identity, and this national identify is explicitly mediated by Qatar Biobank and by national plans for improving citizen health through personalized medical therapies, obstetrics, and treatment for both inheritable diseases and "lifestyle diseases," such as obesity and diabetes. Qatar's planned biomedical development is contributing to the material infrastructural development of Qatar as a player in the global biotech and biomedical research arena while simultaneously promising collective "national" development and biological improvement of the population. I thus show how biomedical development in Qatar can illuminate the particular nature of citizen-state relations, the emergent Qatari national identity, and the metaphysical and infrastructural nation-building projects of Qatar.

In chapter 5, I look more closely at the Qatar Genome Programme and issues of ethics in relation to Islamic law. I show how the Qatar Genome Programme has revealed diverse origins of the Qatari population, and I discuss the possible consequences of these findings for tribal identities in the region. The key findings of this chapter are that the Qatari nation is not a clear-cut genetically homogeneous entity and that genomic research on ethnic populations in the region raises several problems in Islamic ethics, particularly in relation to protecting privacy and in recognizing the economic value of genomic data.

In chapter 6, I conclude with a summary comparison of my research in Israel and Qatar, and I discuss the relevance of my findings to the social theory of science and technology. I argue that the phenomenon of "genomic citizenship" that I describe rests on a hybrid ontological scheme, blending the production of scientific knowledge in the naturalist mode of identification with nationalist imaginaries in the analogistic mode of identification. The nation cannot be defined as a purely natural or biological entity. Rather, the nation is reified, reinscribed and refracted through genomics research and discourse.

ACKNOWLEDGMENTS

This book is a revision of a dissertation that was shepherded by three exceptional senior scholars. I am hugely privileged to have had the supervision of Steve Caton, Jean Comaroff, and Sheila Jasanoff. I have benefited enormously from their extensive knowledge and experience, and the traces of their influence can be easily identified in this text and between the lines.

Acknowledgment of course must go to the institutional support that made this venture possible. Harvard University's Graduate School of Arts and Sciences (GSAS) funded my studentship at Harvard; the Israel Institute awarded me a postdoctoral fellowship for my year of fieldwork in Israel; and the Department of Anthropology and the Center for Middle Eastern Studies (CMES) at Harvard University, my intellectual homes and primary affiliations during this time, funded shorter trips to Israel and Qatar with awards from the Teschmacher Fund and the Arabian Peninsula Research Fund. The Program on Science, Technology, and Society at the John F. Kennedy School of Government has also been an intellectual home during this period. Follow-up research in Qatar and Israel was also supported by a start-up grant at Nanyang Technological University.

Many thanks to Professor Granara at CMES for supporting my trips to Doha and sending me for a summer of Arabic lessons in Muscat. Appreciation to Patrick O'Brien and Emily Burns at GSAS for their consideration and assistance when I requested leave to do fieldwork. Thanks to Michael Koplow and the Israel Institute for facilitating the deferral of a generous fellowship to allow me to complete my coursework at Harvard.

Without this flexible support, long-term fieldwork in Tel Aviv would not have been possible. Particular thanks to Marianne Fritz in the Department of Anthropology, and to Harry Bastermajian, Elizabeth Hope Shlala, and Carol Ann Lister in CMES, for their support, help, and advice.

This project began while I was still a student at the University of Chicago. Thanks to Ariela Finkelstein who in the roasting Chicago summer of 2012 patiently helped me begin learning Hebrew. Thank you also for introducing me to Liran Yadgar, then a student at Chicago, who is now a dear friend, and from whom I have learned a lot about the Hebrew language and Jewish culture.

Special thanks to Irit and Osnat Aharony at Harvard's Modern Hebrew Language Program, for the many hours of instruction, conversation, and kind hospitality. To Lauren Herman, whom I met in Hebrew class at Harvard, it was a pleasure working with you on our class project, and later writing our article together.

During 2013–2014 I enjoyed the experience of a visiting fellowship at Harvard's STS Program. Thanks to Sheila Jasanoff for initially welcoming me as a visiting fellow, and for continually supporting my endeavors. As well as gaining excellent training in STS, I have learned a lot from Sheila about leadership, organization, and program building. For this, I am very grateful. Thanks also to Shana Ashar, at the Harvard STS Program, for help with conference organizing and administration.

The Edmond J. Safra Center for Ethics at Tel Aviv University welcomed me in Israel, where I spent 2015–2016 as a visiting postdoctoral fellow. Many thanks to Shai Lavi and Hagai Boas for inviting me to the center. It was a fantastic year of stimulating conversation. Thanks also to the center's senior fellows, José Brunner, David Heyd, and Haim Hazan, for their intellectual leadership and scholarly example.

I owe special thanks to David Gurwitz and Noam Shomron at the Sackler School of Medicine at Tel Aviv University, who brought me into their laboratories as a visiting scientist. Only with their openness and support was this research possible. Thanks also to all the members of Noam's lab, who welcomed me and made space for me at the bench. Special mention to Keren Oved for accepting me as a colleague with friendliness and openness. A nod also to Mirko Garasic for good friendship and company in Tel Aviv, especially when times were hard.

My initial integration into Israeli life was greatly facilitated by the hospitality of my hosts Ari and Monique, with their beautiful dogs, Qimmiq and Maverick. We have since become very good friends. Thank you for your company, advice, for the many wonderful days we spent together traveling around Israel, and for the evenings we shared over long meals. I look forward to many more.

I am grateful to Brandeis University's Summer Institute for Israel Studies, and the Schusterman family for funding the program that accepted me in 2015 and brought me on a study tour of Israel. This tour introduced me to the tremendous diversity of Israel. There could have been no better way to launch my fieldwork in Israel. Thanks to Ilan and Carol Troen in Omer, for their kindness, and for taking the time to help me find my way in Israel.

To Khalid Fakhro and Mohamed Ghaly in Doha, thank you both for taking the time to share meals with me and for generously explaining the Qatar Genome Programme to me.

I would like to thank Fawaz Mansour at Tel Aviv University, and Nasr, Muhanad, and the staff at CIL Muscat, for helping me get to grips with Arabic.

Toward the end of the project, Molly Mullin provided thoughtful editorial advice that greatly improved the manuscript.

To Porat and Gidi, who hosted me many times both in Tel Aviv and in Newton, Massachusetts. Thank you for your hospitality and friendship. I'm honored you invited me to your home, and I'm very grateful for your encouragement and wisdom.

To all of the friends, colleagues, and supporters who shared the journey with me, over meals, cups of coffee, and long evenings of conversation, thank you for the company and for the encouragement.

1 IDENTITY IN THE AGE OF GENOMICS

"Do you have any other passports with you today?" asked the El Al security officer. Had she asked if I were a dual citizen, or if I were traveling on more than one passport, I could have evaded scrutiny by saying no, but the truth was that I had brought my second "Gulf" passport. "Yes, I have two Irish passports, one that I use for travel to Israel and one that I use only for travel to the Arab Gulf states . . . for my academic research," I replied. She directed me to a podium where an older security officer—tanned, bald, and muscular, with twin handguns bulging behind his fitted navy blazer—began asking me questions.

Identities can be multiple, identities can be exclusive, and identifications can exclude. In the Middle East, perhaps more than elsewhere, it can be difficult to be friends with everyone. You may have to pick a side, present an identity, and make your loyalties visible. Traveling between the Arab Gulf and Israel, for example, presents difficulties, especially when your passports contain stamps of a country that other states don't recognize. The traditional role of the anthropologist as a relatively impartial outsider comes under strain when traveling between quasi-enemy states. Practically, at least in my case, this meant that evidence of my past travel to Arab states made travel to Israel difficult. Reciprocally, travel to some Arab states is impossible with an Israeli visa or stamp.

By the time I decided that I was going to include Qatar in my study, I had already been living in Israel for several months, so I decided to go to the Irish embassy in Tel Aviv and request a duplicate passport. A "clean" passport would facilitate safe and smooth passage to the Gulf, where I needed to travel for my research. I submitted my application and, after a few weeks of waiting, I collected my duplicate Irish passport from the embassy. After getting that second passport, I used it to travel, and I had no difficulty traveling through Dubai, Doha, or Muscat during the following year. At that point, I believed that an unmarked Irish passport would be enough to make trips to both Israel and the Gulf. But there were nonetheless difficulties.

After I had completed my year in Israel and settled back into life in Massachusetts, I returned to Israel for a two-day retreat at the Edmond J. Safra Center for Ethics, in the Judean Hills outside Jerusalem. Outgoing holders of fellowships at the center, including myself, would present research, and incoming fellows would introduce themselves and become familiar with the center's goals and practices. At the airport in Boston, I took my place in the security line that precedes check-in for El Al flights. As the national airline of Israel, El Al has heightened security, more thorough than any I have experienced in my travels. While Israeli passport holders are swiftly ushered to the check-in desks, travelers of other nationalities may face anything from a few security questions to a sustained interview.

My questioning lasted more than an hour, and centered mostly on the summer months I had spent in Muscat: Had I been in touch with locals? Had I been asked to convert to Islam? Was I still in contact with anyone there? After about twenty minutes, the questions turned to my activities in Israel, whom I knew there, and why I was returning. At one point the agent asked if I had learned Hebrew, and when I said yes, we switched languages and continued the interview in Hebrew. During this time, I noticed, by chance, that my Hebrew teacher from Harvard, Osnat, was approaching the security check. She was dropping her son off at the airport. When a security guard asked me where I learned Hebrew, I pointed out Osnat as my teacher. The security guard shuffled over and had a brief conversation with Osnat (who, I subsequently learned, personally vouched for my character), and then he sent me on my way to the ticket desk.

The gate to the flight was about to close, and the security staff were rushing to process the remaining passengers as quickly as possible. The female guard directed me to the side again, where an Israeli security agent performed a rushed, but meticulous, search-and-swipe analysis of my hand luggage.

When I was finally approved to proceed, the security agent prepared to apply an orange security sticker to my passport. Before he could attach it, another guard abruptly intervened, shoving two American passports into the agent's palm, telling him, in Hebrew, "two more Americans . . . Jews, however." Up to this point I had accepted that the security procedures I was being subjected to were reasonable and beneficial for the security of all the passengers, but this intervention, and the justification that the two remaining American passengers should supersede me in the queue for approval because they were "Jews, however," struck me as illiberal, unseemly, and offensive. Further, the insult seemed underhanded, as I presumed he thought that I didn't understand Hebrew. Though these two passengers were officially American citizens, their Jewish identity gave them priority in terms of security and service.

Unmarked passports carry identities beyond their legible inscriptions. Invisible identities matter. I include this anecdote as an example of how the "ethnos" slips into daily life, thought, and practice. Much like how the American passports slipped into the guard's hand ahead of mine, ethnic identity slips into daily consciousness in Middle Eastern ethnonations. It is this "Jews, however" moment that I want to hold in mind when approaching the topic of ethnicity and the entailed prejudice, discrimination, and solidarity that inclusion affords. The phenomenon of ethnic privilege is nothing new, nor is the idea that ethnicity might have something to do with biology. But now, rapidly emerging genetic technologies facilitate a heightened attention to ethnic origins. Genetics research and the application of genetic knowledge of ethnic populations are consequently impacting ethnic identities and ways of imagining inclusion in the ethnos. Questions arise: How do genetic markers become a category used to imagine national inclusion? How is medical research on ethnic populations in the Middle East, or anywhere, driving the imagination of the ethnos as a natural entity? And how does knowledge stemming from that research affect how ethnicity is understood by individuals and by governments attempting to regulate populations? These are important questions as we face the possibility of genetics being used to determine citizenship rights.

Indeed, the governments of Israel (Zeiger 2013) and Kuwait (Cook 2016) have announced that in the future they may use genetic tests of the whole population of citizens, either to determine who has the right to become a citizen or to catalog citizens in a genetic registry, a possibility that forms part of what I call "the molecularization of identity"—following Nadia Abu El-Haj's (2007a) use of the term "the molecularization of race" (see also McGonigle and Benjamin 2016)—to think more broadly about ethnic and national identities in Israel and Qatar. But before I tell the story of the molecularization of identity in Israel and Qatar, I need to outline the state of genetics research, so as to correctly situate my analysis in the context of global genetics in the age of "biocapital" (Rajan 2006).

BIOPOLITICAL FUTURES AND PERSONALIZED MEDICINE

At a TEDx talk in Tel Aviv, Noam Shomron, a leader in the field of genomics, introduces his topic by asking his audience, "What if I told you I could read your DNA, your genetic makeup, your book of life? . . . Are you interested? . . . I can give you a vast amount of information about yourselves that will help you lead better and longer lives, isn't that wonderful? Do you want to?" (Shomron 2017a). After exaggerating the promises of genomics research, he delivers a more modest review of the potential for better medical treatments. He describes his lab as a "Genomic Intelligence Team" that has been sequencing the DNA of hundreds of patients, identifying the exact mutations that cause their medical conditions. Despite his opening lines, he offers his audience more than a sales pitch. He considers what genomic technologies may mean for society at large when, he says, "We are reaching a time we call 'DNA of everything.' We will be able to read the DNA of everything around us." Should individuals with mutations that render them susceptible to infection be banished from schools? Should children who might develop a disease later in life be aborted? Should politicians have to disclose their genetic information to ensure that they will remain healthy during their tenure? In his lecture, Shomron describes "responsible genomics" as giving people the "right information at the right time." To do this, he explains that his lab is dividing genetic information into "private parts" and "public parts" and ranking genes based on their risk in terms of potential disease.

Genomics appears so promising and powerful for many reasons, including its use of complex computational analyses. It is fast ushering in new regimes of health care. Scientists, clinicians, and policymakers imagine that soon individuals will be treated with personalized, precision therapies tailored to their particular genetic and medical profiles. For example, patients might soon be prescribed specific antidepressants based on their genetic profile so as to increase the chance of efficacy. And you may be prescribed a specific drug to lower your cholesterol based on your genetic profile. In the near future, therapies will be designed and customized to fit each patient better, or so the story goes. These technological possibilities afford real hope for improvements in the treatment and prevention of some illnesses. However, they also raise biopolitical and ethical concerns while simultaneously promising a utopian healthy future.

Precision or personalized medicine denotes emerging medical models that use molecular diagnostics, genetic sequencing, cellular analysis, and pharmacogenomics to tailor individual health-care treatment and prevention. By taking genetic, environmental, and lifestyle factors into account, and by relying heavily on big data analysis, precision medicine aims to identify risk factors and biomarkers that predict health outcomes and the best treatment for the patient. Recent advances in the speed and efficiency of genetic sequencing technologies mean that clinicians will very soon be able to quickly and cheaply obtain the full genomic sequence of their patients and learn about the role of specific genes for each individual. Genomics is thus becoming discursively constructed as central to the development of an effective system of "personalized" medicine.

Among STS scholars you can find deep skepticism about the view that genomics holds great promise for humanity. Shomron and the scientists I met believed in that promise. But I did not find them to be naïve technocrats. I did not consider them to be driving their science forward because of their professional, commercial, or nationalistic agendas. They typically are acutely aware of ethical concerns and their responsibilities, and they often contribute to public debates on these matters.

However, genomics research takes place in laboratories and institutions with unique characteristics in each location. The genomic technologies that precision medical models rely on have also been used to describe the genetic structure of particular regional and national populations, thereby making genomics an engine for driving visions of a generic utopian future based on technical progress. They also can encourage and naturalize biological understandings of ethnonational or racial communities. The general global development of genomics entails unique particularities in distinct contexts.

Such a confluence of national imaginaries and global promissory futures is powerfully epitomized by the phenomenon of the "national biobank," a repository of human tissue and information from citizens, which entangles the unique health concerns and the sense of collectivity of a single national community while also contributing materially and metaphysically to the global progression toward personalized medicine. The citizen becomes both consumer and "patient in waiting" (Rajan 2006).

At the time of writing, national biobanks have been established in Iceland, Ireland, Canada, Australia, Japan, Singapore, Kuwait, Israel, Thailand, Belgium, Luxembourg, Estonia, South Korea, Dubai, and Qatar. Biobanks offer a window into the ethnopolitics of biomedicine and the national communities that incubate them and endow them with symbolic power as repositories. The future practical success of precision medicine depends on the establishment of large biobanks that collate diverse data, including family genealogies, disease histories, drug sensitivities, and genomic data. While these initiatives hold promise, they also raise social and ethical challenges, specifically regarding the enrollment of volunteers in large genetic databases; the need for a transformation in the mind-set of clinicians, patients, and the broader public; and the need for ethics informed by multiple disciplines to address legal rights and responsibilities, particularly with respect to personal privacy. In other words, the future potential of "personalized" medicine crucially depends on the "collective" participation of informed citizens and a wide range of stakeholders. So far, nation-states have been key actors in making this possible.

In 2012, UK prime minister David Cameron launched a five-year, £300-million initiative to sequence 100,000 genomes from UK National Health Service patients with rare disorders, cancer, and infectious diseases (Marx 2015). Similarly, in early 2015 President Barack Obama announced a $215 million effort to couple patients' physiological and genetic data and improve the "precision" of individual treatment (White House 2015; see also Reardon 2015). The Chinese government followed suit in March 2016, launching the "China Precision Medicine Initiative," a 15-year, $9.2 billion plan to establish the country as a global leader in precision medicine (Perez 2017). Comparable projects are also under way in Australia, Japan, Canada, Singapore, Kuwait, Qatar, Israel, Thailand, Belgium, Luxembourg, Estonia, and South Korea. The medical benefits entailed by these ventures could be significant, but the impact on the way health care is practiced and how citizenship is performed remains unclear. Further, "precision medicine," it has been argued, "is much more than just genetics" (Lewis 2015). While it is thought that precision medicine will lead to the "prevention" of many diseases, such data-gathering efforts will also likely lead to new therapeutic strategies, entailing new ways of thinking about the role and experience of the patient. In the future, for example, you may be considered "sick" based on your genetic risk long before you experience any illness. This will change the limits of disease experience, such as what counts as healthy or unhealthy and at what point medical intervention is recommended. Likewise, the way diseases are categorized will also change (European Science Foundation 2012; National Academy of Sciences 2011). For example, in the future diseases might become taxonomized on the basis of the underlying genes or variants, rather than on a similarity of symptoms.

The coming era of precision medicine now depends less on technical and scientific advances than on ethical and sociopolitical developments. To obtain meaningful and statistically significant genetic readings of patients and then implement useful pharmacogenomics databases, you first need the voluntary and informed participation of healthy populations. You need large-scale genetic database projects before you can match individual molecular-genetic readings with clinical diagnostics. This, in turn, could reveal how the diverse genetic makeup of populations relates to individuals' varying responses to treatments. Because of this possibility, massive databases will probably be established, collating family genealogies, disease histories, drug sensitivities, and genomic data in an integrated system. When I interviewed Noam Shomron in 2016, he told me that in Israel it is increasingly common that, when a family member is ill, healthy relatives are also sequenced or may be asked to act as treatment "controls," to help identify the pertinent genetic factors. To make precision medicine work better, however, quality long-term medical records and oral family medical histories will also be essential.

For health care to shift from a focus on treatment to one on prevention, however, a major change is required in the mind-set of clinicians, scientists, patients, close family members, and members of the health-care industry in general. TEVA Pharmaceuticals, a large Israeli pharmaceutical company, has stated that the major challenge facing personalized medicine is, in fact, the "reversion of healthcare from treatment to prevention" (Shomron 2014). TEVA pointed to the potential of next-generation sequencing as promising a major boost for the development of personalized medicine but emphasized that health-care providers still need to embrace the "idea" that genetic information is an important part of medical treatment. Progress thus depends on a diverse set of actors. The unfolding of personalized medicine and the building of large-scale databases with the collective and voluntary participation of both patients and healthy citizens depends precisely on such a change in mind-set. Health identities must be reconfigured. Novel sociotechnical imaginaries must be consolidated. New populations may be brought into being.

Moreover, the kinds of collaborations among clinicians, patients, scientists, and the broader public that drive these developments will likely change, as patients' and health professionals' roles evolve. Furthermore, wider public engagement in debate and decision-making could further public engagement in what has been called good "citizen science" (Prainsack 2014a). The benefits such databases will provide may still be unknown and are perhaps, at this point, inestimable, but some of the problems that such collective projects raise are already very clear: the issue of genetic privacy, the ethics of data sharing, the effect on health insurance, the rise of medical "risk" status, and the psychological effects on people and close kin, particularly if they are informed that they carry a pernicious risk factor.

Genetic databases raise important questions of ethics and identity. For example: What is the social nature of the "individual" person in relation to their social community (Prainsack 2014b)? Who owns genetic data? What are the risks of sharing family data? What will be the negative impact of unearthing latent, but potentially negative, genetic data? Will the human genome be broken into regions, or novel "families" of genes, weighted differentially and dynamically according to their known significance? And, what are the legal or "bioconstitutional" (Jasanoff 2011, 3) provisions for participants who may wish to withdraw their personal medical-genetic data, or that of their relatives, after initial consent (Gurwitz 2015)? Biobanks, however, are not an unprecedented phenomenon in terms of the sharing of biological material. There are many suitable comparisons to aid the anthropological analysis of genetic databases for personalized medicine. In Israel, a voluntary blood donation system has been established by the National Transplant Center.[1] Under this plan, individuals who elect to donate blood receive a government identity card assigning them priority in the event that they ever need emergency blood donations (Magen David Adom in Israel 2015). A similar system in Israel exists for organ donation, called the "Adi card."[2] To understand how genetic databases might engage the public, it should be helpful to understand what makes such systems possible and whether they are effective. There also may be unique challenges to fostering public cooperation with genetic databases.

Individual patients and citizens may object to sharing their genetic data, perhaps out of skepticism or fear of the impending changes in the way medicine is practiced. For example, with the advent of personal genomics we can only expect a rise in the number of identifiable "risk factors" and choices of prophylactic medication. One legitimate public concern is that the emerging logic of "prevention by treatment" could go too far, costs would spiral out of control, and whole populations could become overmedicated for "risk" (Rajan 2006), with millions of people being put on multiple prescriptions for the rest of their lives (Dumit 2012). That said, personalized medicine still holds real promise, especially for rare genetic "orphan" diseases, which have generally been neglected by the mainstream pharmaceutical industry and which need and deserve more attention in order to deliver a parity of care to sufferers. Furthermore, as people are now living longer, the impact of neurodegenerative and autoimmune diseases (so-called diseases of aging) will only become greater. In this regard, genetic predispositions for late-onset diseases will also become more important as we enter the age of risk and prevention. The issue of genetic markers for disease will become clear in the research I describe taking place in the lab in Tel Aviv.

GENETIC SELVES

Technologies for sequencing and storing human genomic data, and for analyzing genetic information, are rapidly increasing in speed and power. Tasks that used to take months can now be performed in seconds. It has been estimated that "1 in 25 American adults now have access to personal genetic data" (Regalado 2018). These developments necessitate appropriate governance and ethical policies so that individuals and groups can be sufficiently informed about what is at stake, so they can protect their genetic privacy accordingly. A major challenge, and an important ethical consideration in the development of personalized medical models, however, is the establishment of databases that couple genetic and phenotypic data (clinical information about the person), which at this point are considered "sensitive." Biobanks and databases thus raise social and ethical questions, such as to how to protect the genetic privacy of volunteers consenting to the use of their personal sensitive data. Moreover, the significance of those data may change as technologies and analytic capacities increase in power, making it important to have long-term security measures in place. And genetic data may hold different levels of importance in different contexts. Some communities have a low threshold for sharing their clinical and genetic data. Some patients might be eager to share their data, in hopes of furthering particular areas of medical research, especially in cases of rare genetic diseases for which personal family history is involved. But as technologies change and the amount of personal biomedical data increases, so do concerns about genetic privacy.

If sequencing, for example, moves from sequencing DNA to sequencing RNA (a molecule related to DNA with a wide range of biological roles), a likely future development, this would require reevaluating the information yielded from each technology and also the degree of vulnerability of each dataset. RNA, being a transient messenger molecule that can vary by tissue type or health status, may hold less information about the disease risk you may have inherited from your parents or passed on to your children. In addition to considerations of risk, public trust in medical institutions is crucial if the science is to proceed. Trust, however, is needed not only between researchers and participants but also between governing bodies and scientific communities. Already there is a rise in doctors, genetic counselors, and for-profit companies interested in genomic data. Moreover, patients may arrive at a clinic with more genomic knowledge than the researcher is seeking to investigate (they may have sequenced their genome already), raising the question of the balance of information and power between researcher and participant.

Biobanks and genetic databases also impinge on the configuration of privacy, the way in which the national collective is understood, and, crucially, on the ontology of the "in-dividual" self. In terms of the social nature of genetic data, consider the proposition that genetic data might be considered "individual personal property," which can be legally protected as such. Since humans (usually) gain their genetic signature through biological inheritance from two parents, which they share with siblings, much information about an individual can be extrapolated by examining close relatives' genetic data. The fact that genetic data, or metadata, could be easily acquired by investigating a person-in-question's close relatives challenges the notion that genetic data are "individual" in any categorical way. Rather, personal genetic data are precisely "dividual" in nature. Dividuality is also an anthropological concept from the study of kinship that describes the intersubjective nature of personhood in contextualized social relations. The dividual self is a distributed entity, relationally constructed, partible, composite, and essentially divisible (Gell 1998; Mosko 2015; Strathern 1988; Wagner 1991). In relation to genetic personhood and notions of the limits of personal privacy, human genetic personhood and identity might be better considered as being "dividual," rather than individual, in the sense that genetic data are usually partially shared with close kin, who may also share relevant family, health, and life experiences. As we will see, it is precisely the "dividual" character of genetic data that furnishes the possibility of imagining the nation as a shared biological entity and indeed fosters the imagination of genomic citizenship.

The fact that genetic data and the associated personal medical data are precisely "dividual" in nature may also influence ethical standards, legislation, and governance structures. Legal "individual" citizens will have to recognize that when they disclose their perceived personal genetic data publicly, they inadvertently also share data about their biological kin. This fact becomes more readily apparent with the use of genomics in forensics. In 2018, the so-called Golden State Killer was arrested in the United States after detectives tracked the suspect down through genetic analysis. The police had previously linked the Golden State Killer to more than 50 rapes and 12 murders from 1976 to 1986 but the investigation had gone cold for decades. By uploading a DNA sample collected at one of the crime scenes to a recreational genetic ancestry website (Kolata and Murphy 2018), the suspect was tracked down based on genetic relatedness to participants who had shared their DNA with the genealogy service. In this instance, genetic databases were used for the public good of bringing a notorious murderer to justice, but in the field of health care, there are other bioethical issues to consider when sharing personal genetic data.

The disclosure of your individual genetic data may entail damages to related individuals who could suffer discrimination as a consequence, particularly if you share an inheritable disease status. This potentiality raises more questions about collective consent, the responsibility to disclose or restrict data, and the limits of personal and family privacy. But a concern for protecting individual privacy conflicts with the growing economic valuation of data and the demand for collective databases.

Personalized precision medicine will soon be a viable option for many patients. With advances in the speed and ease of complete genomic sequencing, and in-the-clinic sequencing of other molecules and states (RNA, chemical modification of DNA, and more), it will likely be possible to make better diagnoses and design more effective, tailored treatments for patients. The development of routine genomic tests in clinical diagnostics will affect the commercial value of personal genetic data. Much like the targeted advertising common on social media today, in the future individuals with certain genetic markers could be identified and contacted in advance as a potential customer for certain drugs or therapies. This kind of tailored, targeted treatment could improve outcomes and prolong healthy life. It also could entail more focused direct-to-consumer marketing of medications and therapies, particularly in relation to long-term prophylactics, like drugs for treating lifestyle diseases (such as hypertension, hypercholesterolemia, or obesity), or indeed common psychiatric drugs (like antidepressants, anxiolytics, and drugs for attention deficit disorder). Populations that have volunteered their data could quickly become potential customers to pharmaceutical companies, and as such, individual citizens may want to be able to restrict the access other agents have to their genetic data. On the other hand, as bearers of valued data, individuals may also wish to capitalize on that data. By sharing their genetic data with insurance companies, for example, individuals could potentially benefit from being deemed lower risk for some conditions, and may perhaps even benefit from a lower insurance premium. All this means that the individual self could become a multitude of probabilistic data sets, which overlap with those of biological kin and a broader national cohort. In this way, national citizenship becomes entangled with biomedical research and clinical treatment.

Citizens could also become a new kind of biological citizen-consumer, extracting value from their personal data. But since national publics have typically been paying billions of dollars in annual health insurance policies or through state health programs, thus allowing the companies to become large and influential, it is arguably the responsibility of the companies to reciprocate and pay back something to the community, perhaps by sponsoring data-sharing initiatives. This could be seen as a proactive step to prevent diseases, to aid early detection, and to categorize patients at risk and carefully monitor their health. In fact, Human Longevity, an insurance company in South Africa and the United Kingdom, made the pioneering move of offering subsidized genetic tests to its policyholders, as long as they opted to adopt health improvements that can defer potential sickness, like stopping smoking (Human Longevity 2018). Incentives like this might help insurance companies and their customers begin to work together to improve human health outcomes and to lower disease risks. Such potentials in data sharing and personal genetic medicine will probably lead to the development of algorithmic systems that can measure the relative value of the data relating to specific genes, groups of genes, or RNA. With the tremendous market value that is created by these sequences and their complex relations, it becomes all the more important to understand how to protect privacy and anonymity in genomic research.

THE INTERFACE OF ANTHROPOLOGY AND STS

Pierre Bourdieu devoted his final lecture at the Collège de France to the subject of science, because he believed that the "world of science is threatened by serious regression" (2001, vii). He argued that "the autonomy that science had gradually won against the religious, political or even economic powers, and partially at least, against the state bureaucracies which ensured the minimum conditions for its independence, has been greatly weakened" (vii). He added that "the boundary, which has long been blurred, between fundamental research, in university laboratories, and applied research, is tending to disappear completely" (vii). These lines may seem somewhat outdated. In STS and the anthropology of science, it is now widely considered more productive to think in terms of a dialectical, mutually constitutive relationship between the applied and the pure, rather than dwelling on boundaries. Bourdieu's comments nonetheless capture something important about the changing politics of science. In fact, the way in which practical usefulness is concealed in basic research pursuits remains part of the indigenous conceit of "native science." My concern is that precision medicine, as outlined above, masks the potential utility of genomics in state-led population management, and indeed the reification of ethnic and national groups.

Scholars in the social study of science have described how scientific knowledge is influenced in complex ways by the historical, social, cultural, and political climate that incubates it (e.g., Abu El-Haj 2001; Bijker, Hughes, and Pinch 1987; Bloor 1991; Daston 2000; Franklin 1995; Jasanoff 2004; Jasanoff et al. 1994). Science, according to this body of scholarship, may appear a universal practice, but it varies historically and across nation-states. Social order and scientific knowledge are coproduced in complex entanglements that cannot be neatly separated into the analytics of pure and applied, or "nature," "culture," and "politics" (Cooper 2008; Hogle 1999; Jasanoff

2005; Latour 2004). Moreover, new technologies can give rise to new populations: natural biological populations become political populations, indeed hybrid biopolitical populations, as research "loops" back and impacts the populations studied. Navon and Eyal (2016) call this phenomenon "looping," building on the concept of "dynamic nominalism" developed by Hacking (1995, 1998, 2006) to describe the way the study of autism as a disease constitutes autism populations as a new entity. The important point for this book is that medical research on specific populations can transform the self-perception of those populations and help generate new social identities based on their newfound natural character.

Novel methods of classifying human populations are emerging in medical research so that in biomedical research, the middle-class, middle-aged white male is no longer the basis for extrapolating normal biomedical parameters of the wider population. Gender, race, and ethnicity have emerged as important categories in the evaluation of diverse populations. In his book *Inclusion: The Politics of Difference in Medical Research*, Epstein tells "a story about the politics of how human beings are known, classified, administered, and treated" (2007, 17). Today, certain drugs may be more effective in males than they are in females, and different ethnic groups may have a different range of responses to the same drug. For example, "In 2005, the U.S. Food and Drug Administration (FDA) licensed a pharmaceutical drug called BiDil for treatment of heart failure in African American patients only. Having failed to demonstrate the drug's efficacy in the overall population, BiDil's manufacturers reinvented it as an "ethnic drug" and tested it only on African Americans" (2).

Social identities intersect with medical research and access to appropriate therapies, rendering the right to health a domain of identity politics so that today, according to Epstein, "we are witnessing a repudiation of so-called one-size-fits-all medicine in favor of group specificity" (2007, 5). The new "policies of inclusion" connect a way of performing a biological citizenship to ongoing debates and discussions of civil rights, responsibilities, and concerns about heritable diseases that are unevenly distributed across different groups. More recently in the United States, for example, the research program All of Us was established to help better include populations that have been underrepresented in biological research. All of Us aims "to create a research database that reflects the diversity of the U.S." (Precision Medicine World Conference 2018). Dr. Stephanie Devaney, research program deputy director at All of Us, states that "achieving a demographically, geographically, and medically diverse participant community is a top priority for us. . . . A diverse participant community will fill gaps in our scientific knowledge and give everyone the chance to benefit from biomedical research." One of the key points Epstein (2007) makes, however, is that we shouldn't necessarily assume that "inclusion" means better and fairer. Even as human diversity resists unproblematic categorization into ethnic and medical populations, different nations engage in distinct forms of scientific reasoning and persuasion with different, idiomatically particular understandings of transparency and trustfulness underpinning their practices and rationalities. Biomedical research embeds different social and political values in different contexts.

In her study of bioprospecting in Mexico, Cori Hayden outlines the ways in which modern science embodies interests and human values. She reiterates the core principle of the social study of science that "what makes a fact authoritative is not merely its resemblance to 'nature' but rather the robustness of the social interests that can be enrolled in its support" (2003, 21). In this reading, "(scientific) knowledge does not simply represent (in the sense of *depict*) 'nature,' but it also represents (in the *political sense*) the 'social interests' of the people and institutions that have become wrapped up in its production." In her view, the task for science studies is "to "identify, uncover, or reveal the interests that are wrapped up in knowledge and artifacts" (21; emphasis in original).

In my research, in particular, it is the interests and logics of states that come to the fore, because I am concerned with ethnonational identity. Jasanoff's (2005) relevant work at Harvard University's STS program has used the term "civic epistemologies" to describe the systematic ways that different national cultures engage with scientific knowledge and make decisions in the public sphere—making clear that if you want to understand the cultural politics of the state, science provides a rich ground for research. "Public reasoning," she writes, "achieves its standing by meeting entrenched cultural expectations about how knowledge should be made authoritative" (249). Science, she contends, even while it is a global discourse, must be apprehended in its specific cultural, national, and institutional contexts.

Other work in the social study of science, however, has emphasized the essential role of "connections" and hence the importance of tracking the networks of humans and instruments that produce new knowledge (Bijker, Hughes, and Pinch 1987; Latour 2005). These insights about "networking" emerged early on when anthropologists began exploring the scientific laboratory as a site of inquiry (Latour 1987; Latour and Woolgar 1986). Before the 1980s, there was little anthropology "at home," and ethnographers paid scant attention to modern science. As Latour and Woolgar put it,

> Since the turn of the century, scores of men and women have penetrated deep forests, lived in hostile climates, and weathered hostility, boredom, and disease in order to gather the remnants of so-called primitive societies. By contrast to the frequency of these anthropological excursions, relatively few attempts have been made to penetrate the intimacy of life among tribes which are much nearer at hand. (Latour and Woolgar 1986, 17)

Latour and Woolgar attempted to sidestep the distinction between "'social' and 'technical' issues, however closely related these might be said to be" (1986, 32).

As an experiment itself in a new site of ethnographic inquiry, they recorded the daily life practices of a biology research laboratory at the Salk Institute in San Diego. They reported on daily conversations, the writing of internal reports and experimental results, as well as the so-called purified products of the research, which emerge as research articles for wider dissemination. They borrowed metaphors from biology practices, such as "purification," to describe the epistemic practices of the laboratory and how epistemic outputs are disentangled from their social history in messy human-machine networks. Their purpose was to elucidate in a constructivist register the internal laboratory practices that contribute to the "construction" of new facts, sometimes entailing new objects. They did not set out to challenge the validity of the epistemic outputs of the laboratory.

Their purpose was to problematize, philosophically, the separation, indeed purification, of fact from value in modern science.

Latour continued with this theme in his work, and he later defines "Science" as "*the politicization of the sciences through epistemology in order to render ordinary political life impotent through the threat of an incontestable nature*" (2004, 10; emphasis in original). For Latour, in modern societies, epistemology is routinely pitted against politics, humiliating politics into submission and masking the political nature and interests of the truth-bearers. These insights have since become the taken-for-granted operating assumptions of STS.

Scholarship in the history of science has similarly moved to embrace an "anti-epistemological" mode of inquiry, investigating the basis of knowing reality by showing that scientific objects can be meaningfully read as being both partially real and historically produced at the same time, evading a constructivist/realist split (e.g., Hacking 2000, 2002). Daston terms this an "applied metaphysics" approach, advocating the analysis of scientific objects in their historical and political context of production and circulation (2000, 1). She writes, "If pure metaphysics treats the ethereal world of what is always and everywhere from a God's-eye-viewpoint, then applied metaphysics studies the dynamic world of what emerges and disappears from the horizon of working scientists" (Daston 2000, 1). Anthropologist Jonathan Marks mobilized a similar viewpoint in relation to the "natural" facts of race, saying "race" has been "genetically real when geneticists who believe it is real brandish their particular genetic data and statistical analysis, and it is unreal when geneticists who do not believe it is real brandish their genetic data and statistical analysis" (2013, 250). This book builds on these critical insights from the history of science and STS when attending to the social life of science and technology. Many prior works in the social study of science have tended to bracket epistemological issues (such as the relation between ethnic or national context and what counts as true or valuable knowledge) in favor of a global political economic reading that emphasizes inequality, identity politics, lack of access to health care, or indeed overmedication of populations (Cooper 2008; Dumit 2012; Franklin 2007; Greene 2014; Rajan 2006; Reardon 2004, 2011). Such works have appropriately addressed the logics of the global market and how corporate economic interests have shaped the life sciences in the twenty-first century. Broadly speaking, this body of work traces how, since the 1980s, there have been significant developments in molecular biology (gene cloning, the Human Genome Project and the development of biological therapeutics, for example), and that this period also witnessed the advent of neoliberal policies that entailed the recession of the state and the dominance of the market in public services and scientific development itself. This confluence of events has affected the field of the biosciences; it has set the stage to usher in a regime of valuation that hinges on the promissory value of "life itself" in capitalistic terms (Cooper 2008, 3; Fortun 2008). In thinking about the interrelationship of the market economy, as it engulfs the biosciences, and the field of the modern life sciences, Cooper builds on the work of Michel Foucault, contending that the "development of the modern life sciences and classical political economy should be understood as parallel and mutually constitutive events" (2008, 5). In this regard, the biosciences may also be read as a manifestation of the logics of global capital, even while their epistemologies explicitly address issues of life and "basic" "natural" science.

Dumit's (2012) incisive work on the political economy of pharmaceuticals also touches on epistemological issues and points to an emergent medical logic whereby the absence of symptoms no longer defines health. He argues that with the marketization of commercial pharmaceuticals, and a research industry that is seeking to determine risk factors that indicate the likelihood of developing a disease later, populations are being encouraged to take prescription drugs years in advance of potentially developing a disease. He chronicles the way "the very concept of a risk factor was created alongside the innovation of large-scale prospective clinical studies" (4). In his constructivist view, "neither health nor illness are states of being: they are states of knowledge; they are epistemic" (13).

This book, in contrast, does not dwell extensively on the political economy of the global life sciences but instead aims to bridge anthropology, Middle Eastern studies, and the social study of science, and asks how and why genomics and biobanking are becoming key sites for imagining national and ethnic communities in the contemporary Middle East. For this work, the transnational context is, of course, pertinent, and the way the contemporary biosciences reconfigure identities is crucial. As Dumit states, in contemporary biomedicine, "risk is now a subjective present illness: treated as if diseased" (2012, 16). This fact is essential to understanding the logics underpinning national biobanking and the value scientists hope to capture.

The focus of this study, however, is on ethnic and national identity. This inquiry lies orthogonal to the question of whether ethnic groups are real, imagined, or constructed, but instead concerns the conditions of the genomic and biobanking practices that render these imaginations of collectivity important today. Recent history, demographic issues, national politics, and global trends in biological science condition the ways in which ethnicity is attended to. The ethnic context unfolds as a structuring force in basic scientific research. My line of thought is at its core a deferral, or displacement, of ontology, in this case with ethnic genetics, and it resonates with an immanent critique of the natural facts of race and ethnicity. Indeed, it is a timely moment to question the natural facts of race and ethnicity, as they have moved to the molecular realm where they are no longer so easy to challenge or refute with common sense.

THE MOLECULARIZATION OF ETHNICITY AND RACE

In the twentieth century, scholars of ethnicity could be distinguished between those taking a "primordialist" approach and those taking an "instrumentalist" approach. The former presupposes a deep and shared historical experience among members of the ethnic group, while an instrumental analysis treats ethnicity as a process of identification that involves a strategy to extract rights and resources (Barak 2002). Moving beyond a simple binary, however, anthropologists Comaroff and Comaroff (2009) have described how in late capitalism elective ethnic identification is caught up in the logics of the global market, with ethnic authenticity itself more and more appearing as a marketable commodity, or a claim to rights to extract value from a national heritage. Ethnicity has become closely inflected by the logic of global capital. Ancestral claims, authentic belonging, tourism, the commodification of culture, and also techniques of governmentality and nationalist political rhetoric have all been entangled with claims of authentic ethnic identity.

Simultaneously, claims of ethnicity have also moved into the molecular realm (Abu El-Haj 2007a; Comaroff and Comaroff 2009; Fullwiley 2008; Gottweis and Kim 2009; McGonigle and Benjamin 2016; McGonigle and Herman 2015; Rose 2007; Sun 2017; TallBear 2013a, 2013b; Tupasela and Tamminen 2015). Abu El-Haj (2007a) has used the term "the molecularization of race" to characterize the phenomenon. The politics at stake in the new discourse of genetic claims to race and ethnicity have been debated, particularly with regard to the reinscription of older racial categories. Genetic claims to history often rest on a "divinatory logic" that seeks out invisible essences (Palmié 2007), to the point that "postgenomics does seem to be giving race a new lease on life" (Abu El-Haj 2007b, 223). However, the difference with genomic definitions of race, when compared to nineteenth-century biological and physiological measurements, is that "junk DNA" that may have no physical expression or biological effect is now being used to mark racial divisions (Abu El-Haj 2007b). It would seem that the political imaginaries of the present engender the immaterial "new genetics" as a possible and, indeed, preferred, source of data for historical and racial mapping. How this will matter in the contemporary Middle East is not yet clear, however. In any event, the "nature" of race may be profoundly influenced by the politics of the present. As Comaroff and Comaroff write, "it is the *marking* of relations—of identities in opposition to one another—that is 'primordial,' not the substance of those identities" (1992, 51; emphasis in original). This proposition underpins the core argument of this book, that the context of ethnonationalism lays the ground for the way in which ethnicity is becoming attended to in the molecular realm.

This issue has been analyzed at length by anthropologist Nadia Abu El-Haj in her 2012 work *The Genealogical Science,* which lays out the field of "genetic history" as it pertains to Jewish origins. Databases that emerged from the Human Genome Project were used by academics to render accounts of the origins of contemporary populations and to evaluate the plausibility of oral traditions and historical narratives. Genetic narratives thus gained credibility as these data are wielded academically to describe the origins of ethnic groups. At the same time, an emergent market in recreational genetic ancestry testing has bolstered the narrative potential of genetics in relation to ethnic identity. Abu El-Haj calls this phenomenon the field of "anthropological genetics." Fundamentally, the field of anthropological genetics studies "human origins and migration routes out of Africa" and "the genetic diversity of the human species" to map the "genealogies of particular populations" (2012, 3). Her work traces how anthropological genetics has emerged from turn-of-the-century race science that relied on cranial measurements and phenotypic differences. This race science was followed by a population genetics from the 1950s that was based primarily on blood group data, which laid the ground for later work based on DNA sequences (4).

Specifically, in relation to genetics of Jewish populations in the newly founded nation-state of Israel, Abu El-Haj (2012) analyzes the work of Israeli population genetics in the 1950s and 1960s. She reads their genetic studies "as expressing a desire—indeed, a *need*—to find a 'content' for the a priori nationalist belief in the fact of Jewish peoplehood." At stake in their research was the possibility of revealing a Jewish "common origin in ancient Palestine" (4). Now, she argues, is a significant moment in which genetic sciences have considerable rhetorical power as, since the mid-1990s, technological developments have steadily extended the social reach and epistemological authority of genomics (5). Her work moves beyond a conventional history of science, however, and she tracks these scientific developments in relation to the broader social and political context. She writes:

> Inspired by a tradition in the history, philosophy, and sociology of science, I pay careful attention to scientific epistemologies, past and present. But I read scientific epistemology via an anthropological sensibility trained to understand not just the epistemological, social, and political conditions of possibility of scientific work. I am interested in giving an account of the forms of life specific biological disciplines make possible or not. Moreover, in writing about genetic history I do not make a strong distinction between scientific and social practices. (Abu El-Haj 2012, 7)

Following scientific developments in the 1980s, Abu El-Haj shows how population genetics entered a new era as knowledge of mitochondrial

DNA grew, facilitating a greater understanding of "genetic evolution, migration, and genealogy" (2012, 8). These developments rendered DNA as "a historical document" (11) so that today "the anthropological gene and genome are molecular archives" (2). Consequently, DNA is progressively becoming read as "indexes of ancestry and origins" (22).

Abu El-Haj argues that the genetic research on Israeli populations that followed the foundation of the state was "a practice wedded to the work of *imagining* the nation" (2012, 64; emphasis in original). The pressing question was what evidence there might be that "the Jews are a nation with a shared origin in ancient Palestine" (64). She emphasizes the distinction between descent and identity in anthropological genetics: "Descent from a common ancestor does not imply identity. Rather, it implies a presumably decipherable matrix of genealogical relationships 'visible' in genetic polymorphisms" (38). In other words, descent has become a statistical phenomenon that is molded in specific social and historical contexts and that creates the conditions that impose value on genetic readings that reveal historical truths. In distinction to racial categorization based on phenotypic or physiological differences, however, the new anthropological genetics makes use of noncoding DNA, which may not have an essential biological role. Nonetheless, noncoding DNA is useful, as it is argued to be a significant form of evidence for reconstructing origins (22). As objects for group making, DNA markers "do two things at the same time: they differentiate groups and, simultaneously, make no difference at all" (23). In this application, anthropological genetics is more than biological citizenship or biological connectedness between members of an ethnic group; it also engenders efforts "to identify a history within" (28).

This book builds on Abu El-Haj's (2012) study of anthropological genetics in relation to Jewish identity, but unlike her work, which is extensively historical, this book relies more on ethnographic work in the present and brings us into the heart of genomics research and biobanking in contemporary Israel and Qatar. It follows the invitation of her rich analysis of the politics of Jewish genetics to ask whether and how the Middle Eastern ethnonation is being apprehended in genetics research today at Israel and Qatar's national biobanks and contemporary societies.

In the past, genetics research in Israel has not only addressed a shared genetic archive among Jewry, but has also reiterated assumed ethnic distinctions within the Israeli population, such as Ashkenazi (European) and Mizrahi (Middle Eastern). In other words, there has been a tight dialectic of context and choice in biological iterations of Jewish identity in Israel. Or as Comaroff and Comaroff put it, "While ethnicity is the product of specific historical processes, it tends to take on the 'natural' appearance of an autonomous force, a 'principle' capable of determining the course of social life" (1992, 60). How collective identity determines social life hinges, however, on the mediation of the imagination of the wider collectivity, indeed, on the imagination of the nation. We have long understood that the nation needs to be imagined to be made real. What part does genetics play in this process of imagination today?

IMAGINING THE NATION

In 1983, Benedict Anderson revolutionized the study of nationalism when he identified its origins in the shared imaginations of individuals as copresent members of a jointly envisioned nation-state. He demonstrated that technologies such as newsprint media are essential in maintaining these collectively shared imaginations and their entailed performances. Appadurai (1990) extended Anderson's notion of imaginaries to describe the transnational flows of a globalized, technologically advanced, and interconnected world. He characterizes such modern spaces by their jarring "disjunctures," junctions where difference is encountered and where homogeneity is challenged. He splits such "global flows" into five dimensions: ethnoscapes, mediascapes, technoscapes, finanscapes, and ideoscapes. With these categories, which widen the critical range of how assembled collectivities imagine their existence, Appadurai puts forth as a concept "Imagined Worlds," drawing on Anderson (1983), but providing a more "schismatic" and "nonlinear" program for mapping the semiotic flows that constitute the multiplicities of these very real, but also, of course, *imagined*, worlds.

Appadurai's intervention allows us to think more broadly about communities in a multidimensional way: as material, image, practice, aesthetics, or as worlds of abstract ideas. It also allows us to consider science and technology as semiotic mediators of collective identities. Science and technology play a role in reinforcing the imagination of a shared national community, or indeed a global scientific community. Jasanoff and Kim (2013) take this line of thinking a step further, and they provide a theoretical framework for understanding the global politics of science and technology that builds on this literature of imaginaries and their relationship to global flows and circulation. They define "sociotechnical imaginaries" as "collectively held, institutionally stabilized, and publicly performed visions of desirable futures, animated by shared understandings of forms of social life and social order attainable through, and supportive of, advances in science and technology" (2015, 4). With the concept of sociotechnical imaginaries, they urge us to ask how ethical, social, and political commitments are built into national visions of technoscientific development and also how science and technology are used by people to imagine their citizenship, identity, and participation in public life.

Biobanks collect substances (blood, DNA, urine, and other tissue samples) from individuals in the population that are literally shared, and consequently open up a range of possibilities for measuring, cataloging, controlling, imagining, and generating populations, and subpopulations, and at the same time open up a range of medical implications and future treatment possibilities. Biological substances become not only ways of imagining a shared community, but also ways of arranging or assembling in contiguity the shared substance and health of the imagined community and simultaneously, as we will see, of modulating how lives are lived.

Ethnic genetics can also be a site where a population is subject to control. Modern nations typically imagine a past that establishes the grounds for a shared sense of community. Communities draw on narratives and images that bolster claims of shared experience, substance, or a national essence. At moments of crisis or emergence, such imaginations of collective identity take hold. In the Middle East—a region particularly fraught with ethnic and identitarian divides—this process has been prominent. In Israel, reinventions were at work in the state's early years. Abu El-Haj, before starting her research on genetics, studied the politics of archaeology, a form of what she calls "colonial knowledge" (2001, 6). In the context of settler colonialism, she argues, archaeology was used to reformulate a national people: "There emerged, in other words, an elective affinity between archaeology's epistemological and methodological commitments and the cultural politics of the Jewish colonial nation-state-building project as both crystallized in early-to-mid twentieth century Palestine" (16). On the creation of the new Jewish state, historian of Zionism Yael Zerubavel states that "the construction of a myth of origins requires the twofold strategy of emphasizing a new beginning as well as discontinuity with an earlier past" (1995, 43). In other words, the self-fashioning of Israeli-ness involved a redefinition of a Jewish self, a breach with history, and with that a resurrection of a new history that would root Israelis to the land of Israel.

The labor Zionists that dominated the culture of the young state emphasized vigorous physical labor as the basis of a new "Hebrew culture" and gave rise to the spread of the so-called Sabra culture among the first generation of Israelis. The Sabra, a new Israeli type, was thought to be tough on the outside and sweet on the inside, according to the common trope that Almog (2000) documents as akin to a kind of new "secular national religion." In this instance, the body was an important site for performing and imagining the new Hebrew citizen. From the early years of the Israeli state, a fervent anti-diaspora sentiment was rigidly codified into an idealized image of a male Zionist pioneer, imagined as a warrior and a worker, an assiduous and productive member of a healthy society. This ideal is constructed in hyperbolic opposition to the stereotype of diaspora Jews, who were perceived by the first Israelis as feeble, effeminate, and even morbid. The virtual ideal of the Sabra, however, preferred the life of action to the values of scholarship and intellectualism, and he was always willing to sacrifice his individual desires for the greater good of the nation. Katriel's (2004) ethnography of language in Israel and specifically her focus on speech as a site of authenticity in the creation of the "New Hebrew culture" is crucial here. She writes:

> There were two major versions of it: one was the neo-Romantic version inspired by the German youth culture of the turn of the twentieth century and its individualist-humanist ethos, which sought to attain personal redemption through the re-creation of an organic-national community. . . . The other

> version of the New Jew was influenced by Russian pre-revolutionary movements that preached the return to nature and to the simple life via menial, productive work. (Katriel 2004, 19)

Such national imaginaries, and the values that their performances propagate, also play out powerfully in science and medicine. Israeli anthropologist of medicine Meira Weiss (2004) has described the legacy of the Sabra culture of the first generation of Israelis in obstetric medicine, saying that a majority of Israelis "agree that giving birth to a child with a serious impairment is socially wrong" while, in contrast, "geneticists around the world usually regard the decision to abort a deformed fetus as primarily personal" (Weiss 2004, 3).

Science reflects not only widely shared cultural values. Large-scale scientific projects also inculcate a new imagination of territory and belonging. The Israeli state drew on basic engineering and agricultural projects to establish the imagination of a progressive state project. In the 1920s, the land of Israel became territorialized through the material-semiotic electrification of British-ruled Palestine by the marking of the landscape with electricity poles (Shamir 2013). Shamir presents the early electrification projects as a political matter—not simply a consequence of community-building, but a constitutive element of nation-building. A confluence of private capital, technical knowledge, and ideas of progress resulted in an infrastructural development that widened the divide between Jews and Arabs. He argues that "electrical connections participate in processes of group formation; take an active part in the performativity of social asymmetries; shape areas and regions and other spatial formations; and actively assemble, sustain, and enable taken-for-granted categories and dichotomies such as the private and public spheres" (Shamir 2013, 3). Foregrounding the role of the material infrastructural fabric rather than the agents who invested the project with support and interest, Shamir shows how the process of electrification under the auspices of a colonial government produced and affirmed ethnonational distinctions (9).

Similarly, Braverman (2014) describes how the widespread planting of pine trees, which symbolized the Zionist emphasis on healthy growth and agriculture, helped territorialize the land of Israel/Palestine. Today, agriculture is a smaller part of Israel's economy and national self-image. The country has embraced globalization, and this has also produced a new cultural fabric to be understood at the level of global capital. Crucially, however, Israel's move to globalization is also entangled with its national identity. This global context is crucial for understanding how ethnonational identity is produced in the twenty-first century.

Ram (2008) gives an account of the globalization of Israel and offers a new paradigm for understanding the cultural shifts taking place in the country. He describes Israel as "bifurcated into two polar opposites—capitalism versus tribalism, or 'McWorld' versus 'Jihad'—that contradict and abet each other dialectically" (2008, vii), borrowing the idioms of "McWorld" and "Jihad" originally coined by Barber (1992). Israel has experienced progressive high-tech globalization since the 1990s, and Ram uses a Hegelian approach to analyze this specific process of globalization as a "contradictory dynamic totality that conjoins these two negations" (2008, 2). These metaphorical terms of "McWorld" and "Jihad" caricature the process in Israel as "a dialectical struggle between a global, capitalist, civic trend and a local nationalistic religious trend" (6). In other words, in Israel, globalization is characterized by a tension between the global and the local. According to Ram, the process of globalization can lift local processes out of the immediate spatial environment and render them part of an unmarked, delocalized, global order, entailing new markets, new types of actors, new rules and norms, and new strategies of communication (2008, 13). Ram sees this process of globalization as "the new stage of capitalism in the intersection of the twentieth and twenty-first centuries" (17). This process, however, poses a challenge to Zionism, with "two opposite perspectives: a postnationalist perspective, which tilts toward global cosmopolitanism, and a neonationalist perspective, which tilts toward local tribalism" (26).

This dynamic is giving rise to a cosmopolitan post-Zionist culture in Tel Aviv, which is the center of an emergent "creative class" of skilled middle-class workers in the arts, music, science, high technology, and medical research. Ram also sketches three waves of popular culture that have unfolded in the history of modern Israel. The first wave, which began in the 1930s and ended in 1977, was characterized by the dominance of the Labor movement (Ram 2008, 153). The second phase began in 1977 with the electoral upheaval (*Mahapach*), which displaced Labor from power for the first time in Israel's history (153). This event marked the negation of the monolithic statist Labor Party of Ben Gurion that was seen as favoring the European Ashkenazi population over Jews of Middle Eastern backgrounds and Arabs. The victory of the right-leaning Likud Party led to a new right-versus-left political dynamic in the country. The right has been represented by a variety of parties, "including the national-religious party (Mafdal) and the Likud Party and at different points in time other—religious and secular—extreme right-wing parties, such as Tozemt (Juncture), Moledet (Homeland), or Israel Beiteinu (Israel Our Home), and in a somewhat different version also the Shas party" (231). Such right-wing parties incorporate a Jewish national-religious platform while also allowing progressive "marketization" of public services and deregulation of the financial market, thus enabling globalization of the Israeli economy.

The third phase Ram describes occurring in popular culture began in the 1990s and involved "the two-edged feature of postmodern Americanized politics: communalism and commercialism" (2008, 153). This historical evolution poses two contradictory propositions for the future of the state: "transmute into an ethnic Jewish state or to transmute into a liberal state of its citizens" (237). Which turn the society will take is not yet clear, but Ram contends that "since the 1990s Israel has become simultaneously more of a market society and more of a tribal society, more neoliberal and more neofundamentalist, more post-Zionist and more neo-Zionist" (238). The state appears to be able to contain the two sides of the dialectic of "McWorld" and "Jihad" that Ram has described.

The Israeli "national biobank," we will see, both fits and doesn't fit these accounts of nation-building, globalization, and the imagination of a shared bodily essence or practice. The imagination of genetic peoplehood has implications, of course, for the way states may manage their populations. Genetics may become a tool for governing the bio-nation.

BIOPOLITICS AND CITIZENSHIP

Foucault (2010) used the term "biopolitics" to describe the way in which modern democratic states manage, and imagine, their populations at the level of life itself. Biobanking could also foster a new regime of biopolitics at the level of individuals' genes, with potential for governments to catalog the "biological citizen" (Petryna 2013) at the molecular level. While an emerging anthropological literature is investigating the various ways in which citizenship is enacted, performed, imagined, and challenged, particularly in relation to transnationality and migration (e.g., Chu 2010; Lazar 2013; Ong 1999; Petryna 2013), little attention has been paid to genetics as a domain for the performance of citizenship in the Middle East.

In both the Israeli and Qatari contexts, global markets and state-led developments are affecting configurations of nationality and relationships to territory. Inclusion in the nation needs to be symbolically mediated in order for it to be a lived and performed reality. Indeed, one study argued that in Israel "conversion" to Judaism had a powerful "stabilizing" effect for the individual (particularly recent immigrants), by making the convert part of the ethnonational fabric and marking them as an incorporated internal citizen (Kravel-Tovi 2015). It is not yet clear how global markets and mobile workforces will use understandings of genetics and technoscience to channel and transform the local underlying cultural logics of citizenship, modes of performing belonging, and ways of negotiating distance and incorporation within the state. It is likewise unclear whether biobanking developments in Qatar are challenging traditional idioms of tribal allegiance, or whether they are contributing to a biological understanding of historical connectedness with other Qataris.

Let us explore genomic citizenship in the concrete context of each state. We hope to discover how and why the Middle Eastern ethnonation is becoming imagined as a biological entity. The chapters that follow thus consider genomic citizenship in the specific contexts of Israel and Qatar.

2 THE "NATURE" OF ISRAELI CITIZENSHIP

GENE TALK

Yashka is an inexpensive shawarma joint perched on the corner of Dizengoff and Frishman, the urban heart of Tel Aviv. Since I didn't have a proper kitchen in my small one-bedroom apartment, I would often stroll down to Yashka to enjoy a lunch of falafel, shawarma, or shakshuka. I liked sitting there, observing people, and noticing the rotating staff of new Russian *olim* (Jewish immigrants to Israel) clearing tables. One winter afternoon I wandered in, bought a heavy shawarma wrap, and after filling a small bowl with the complimentary pickles and tahini, I sat down in one of the green plastic chairs opposite a man about my age. After a few minutes of eating silently in each other's company, he asked, in Hebrew, if I had seen the football match. When I said no, I hadn't, he asked where I was from, and we began making small talk. He was surprised that I knew Hebrew and asked what I was doing in Israel. I said "research," and after a moment of silence, I elaborated: "I'm studying the way in which genetics relates to Jewish identity . . . for example, how the government might use genetic tests to determine who can immigrate to Israel." He raised a finger and said that he knew about this topic. He has been following the philanthropic efforts of a "big Israeli businessman" who wants to fund research in genetics to show that the Arabs in Israel were Jews who converted to Islam in the past and that, consequently, this would prove that "the occupation is bullshit," meaning that there is no occupation. He proceeded to proudly tell me that he is a "right-winger" and that he was pleased with the recent news scandal: a sting operation in which a right-wing activist (from the Ad Kan organization) infiltrated a left-wing human rights NGO that was attempting to expose human rights abuses in the West Bank. In his opinion, the land belongs to the Jews, and the use of genetics to support those claims ought to align with his political views.

I was intrigued by the way in which he seemingly saw no need to separate politics from epistemology. For him, it was a clear question of orientation, support, and brute force. Scientific truth did not stand outside politics but followed conviction. The absence of the epistemic and professional ideal of objectivity didn't even seem an issue for him. Rather, genetics ought to be used as a rhetorical device to undermine the rights of the Arabs in the region and justify Israel's right to the West Bank. In this formulation, the modern separation of fact and value is irrelevant: politics is driven by commitments and the desire to act, not by putatively disinterested science.

This man's stance on the use of genetics in political action, extreme as it is, speaks to the way in which genetics has infiltrated the Israeli popular imagination as a powerful tool in establishing, policing, imagining, and defending boundaries, identities, and territory. But "gene talk" is not limited to the nationalistic consumers of street food in Tel Aviv. In July 2013, Israel's Prime Minister's Office stated that in the future Russians wishing to make *aliya* (immigrate) to Israel might need to take a DNA test to prove their Jewishness (Zeiger 2013). This statement indicated that the state might use genetic tests to verify a legitimate "biological connection" with a Jewish parent or grandparent.

In 2018, Israeli rabbinical courts began recognizing the results of mitochondrial DNA tests as proof of Jewish heritage (Rabinowitz 2019). Mitochondrial DNA tests offer information about genetic background exclusively from an individual's maternal side, which is in accordance with the matrilineal transmission of Jewish identity. The tests the rabbinical courts began using are based on a comparison of mutations in mitochondrial DNA to databases of other nationalities and ethnic groups (Rabinowitz 2019).

This practice of using genetic tests to determine Jewishness, however, was soon challenged by a petition in Israel's High Court of Justice, "filed by Avigdor Lieberman, Yisrael Beitenu and several private petitioners" against the chief rabbinate and the rabbinical courts. In January 2020, Israel's High Court of Justice ruled in a majority decision against the petition and "found that DNA testing to prove one's Judaism should be allowed" (Rabinowitz 2020). The court also ruled that the "petitioners did not prove that the rabbinate acted in a discriminatory manner." However, the court made a distinction between "reexamination by the rabbinical court of the Judaism of someone who was already recognized and registered as a Jew, and conducting genetic tests to prove one's Judaism." The court also said "the rabbinate must formulate written rules on the issue within a year."

If such DNA tests become rolled out by the state, Israel would be enshrining Jewishness at the level of DNA, rendering "Jewish genes" legally legible and making DNA signatures a determinant of basic rights and citizenship for the first time in its history.

Before getting further into the details of "Jewish genetics" as a discursive field wherein imaginations of citizenship and belonging flourish, it will be helpful to provide some context about the Israeli state. The State of Israel is explicit in defining itself as the homeland of the Jewish people and is thus both ethnoreligious and national in its self-image. The commitment to the Jewish character of the state, however, raises perennial domestic concerns, and frequent moral panics, over who is a Jew, how this can be determined, by what credible authority, and about the exact "nature" or fundamental modality of citizenship in Israel. A genetic test for Jewishness is thus evaluated in this context and would supposedly function as an objective metric of legitimate inclusion in the state, constructing a virtual biological border and providing an unequivocal substrate for calculating ethnic belonging.

Although it is unlikely that genetic tests for Jewishness will become the main criterion for securing Israeli citizenship, the rise of "Jewish genetics," and its circulatory semiotics, exemplified most loudly by the state's announcement (Zeiger 2013), demands an examination of the curious relationships among biology, Jewish identity, and citizenship in Israel. This novel and particular form of governmentality, the management of citizens and populations through "ethnic genetics," needs to be situated within this contingent historical moment as it relates to the political philosophy of Zionism, particularly regarding conceptions of Jewish ethnicity.

As a whole, my approach to the molecularization of identity in the Middle East is more comparative and ethnographic than it is historical, but some historical context is important. This short history of the roots of "Jewish genetics" should clarify why the Israeli state is attempting to understand itself in the present through technoscience. My line of thought in this chapter is a historical anthropology of a concept, "Jewish genetics," with a reading that imposes an immanent critique on the phenomenon of the molecularization of ethnicity in the context of the Jewish ethnonation.

What exists depends on how and why we know it. Rather than regarding ethnic genes as being pure essences "in themselves," a "negative dialectical" critique emphasizes the necessary historical particularities of the mediations of their ontological claims (Adorno 1980 [1966]) and strives for the "negation of reification" (Horkheimer and Adorno 2002 [1947], vii). This approach will also help in thinking comparatively about how and why other states, like Qatar, might similarly draw on genetic technologies in determining rights to citizenship and in imagining the borders of ethnic belonging.

JEWISH ETHNICITY

Judaism is a particularly blurry ethnos. And while clear-cut racial divisions (e.g., Black African, White European, East Asian) are perhaps the ideological construction par excellence, the borders of Jewish ethnicity are being complexified and reformulated with the latest next-generation genomic sequencing technologies.[1] "Nature" becomes more political, more geographically and historically specific, and more culturally particular, as people situated in different national spaces find uses for genomic technologies. In the Israeli context, "ethnic genes" have already entered public discourse, especially because geneticists have been describing the genetic structure and historical migrations of Jewish populations (see Atzmon et al. 2010; Behar et al. 2004, 2006, 2010; Bray et al. 2010; Ostrer 2001; Ostrer and Skorecki 2013). It has been said that such genetic research is contributing to a "'biologization' of Jewish culture and historical narrative" (Egorova 2014, 354), as lay commentators now often turn to DNA evidence as a "rhetorical means for inscribing identities," especially to support "favoured accounts of the origin and historical development of the tested communities" (Egorova 2014, 360).

It makes sense that people appropriate these scientific findings. Jewish population-genetics studies often treat diverse diaspora groups of Jews as related cohorts and often trace genetic data to support the narrative of a line of descent from the ancient tribes of Israel mentioned in the Hebrew Bible. In this regard, "Jewish genetics" reiterates and lends credibility to the Israeli state's founding narrative of return to the Holy Land. So-called Jewish DNA may be read through genomic analysis even when "Jewish genes" are located in areas of "noncoding DNA," that is, from genetic material that probably does not in itself determine a specific physical trait. These so-called Jewish genes may not make a difference at all (phenotypically, at least), and yet they would become vital if they become the legible traces that decide rights to citizenship in Israel. Regardless of the validity or biological importance of such genes to Judaism, at issue is the question of *why* genes are becoming a site for the Israeli state to imagine control of the population.

To understand this potential development, it helps to consider the trajectory of the Israeli state, its commitments to religious law, and how this emergent phenomenon relates to a long history of Jewish political thought and imaginations of Jewish ethnicity. Here, the ethnic composition of Israel is crucial. Despite the ambiguity in the legal, biological, and social "nature" of "Jewish genes" and their intermittent role in the reproduction of Jewish identity, Israel is a country of extraordinary ethnic diversity. Many Jewish immigrants have arrived from Eastern Europe, North Africa, France, India, Latin America, Yemen, Iraq, Ethiopia, the United States, Zimbabwe, South Africa, and the former Soviet Union (FSU), and then there is Israel's Arab minority of close to two million people. And while Jewishness has often been imagined as a biological race—most notably, and to horrific ends, by the Nazis, but also later by Zionists and early Israelis for state-building purposes—the initial origins of the Ashkenazi Jews who began the Zionist movement in turn-of-the-century Europe remain highly debated.

Population analysis by geneticists has led to an unresolved debate over Jewish origins (Abu El-Haj 2012; Elhaik 2012; Kohler 2014). Geneticists have begun to describe the genetic basis for common ancestry of the whole of the Jewish population (Behar et al. 2010), even though the historical claims that are entangled with these scientific studies are still contested. One of the most contentious claims made is that European Jews are descended from converts to Judaism from the Khazar Empire, which covered much of Eastern Europe during the second half of the first century CE (Koestler 1976; Sand 2009; Wheelwright 2013). Some rabbis and several population geneticists instead claim that there is a direct line of descent connecting most European Jews to the biblical land of Israel (Sand 2009).[2] But Israeli historian Shlomo Sand argues, "The Jews have always comprised significant religious communities that appeared and settled in various parts of the world, rather than an *ethnos* that shared a single origin and wandered in a permanent exile" (2009, 22).

Regardless, according to biblical narratives, Jews resided in the Levant for several centuries before the destruction of the Second Temple,[3] and historians broadly agree that European Jews resulted from dispersals of Jews to the north into Europe and the Mediterranean in the early Middle Ages. Following expulsion from Western Europe, in around the thirteenth and fifteenth centuries, Jewish communities expanded eastward to Poland, Lithuania, and Russia. As European Jews have arguably experienced much more persecution and suffered more displacements than Jews living in the Arab world, it is unsurprising that political Zionism emerged in the late nineteenth century almost exclusively as a European Jewish political movement, with the large-scale immigration of Jews from the Arab world not beginning until the foundation of the Israeli state in 1948.

Interest in the topic of Jewish origins is hardly universal among the world's Jews or the communities in which they live. But in Israel, the stakes of the debate over Jewish origins are high, because the founding narrative of the Israeli state is based on exilic "return." If European Jews have descended from converts, the Zionist project can be pejoratively categorized as "settler colonialism" pursued under false assumptions, playing into the hands of Israel's critics and fueling the indignation of the displaced and stateless Palestinian people. The politics of "Jewish genetics" is consequently fierce. But irrespective of philosophical questions of the indexical power or validity of genetic tests for authenticating Jewishness, and indeed the historical basis of a Jewish population "returning" to the Levant, the realpolitik of Jewishness as a measurable biological category could also impinge on access to basic rights and citizenship within Israel. Looking at the issue in the context of Israel's national politics and modes of governmentality, a geneticization of citizenship would mark a new moment in the Zionist political philosophies that motivated the state's emergence, a philosophical set that already varied considerably since many of the European Zionists who founded the Israeli state differed widely on the basic principles on which Jewish nation-building should be pursued.

In connecting genetic identity to nation-building, I follow Weingrod in thinking of nation-building as "processes through which citizens in a society reach broad agreement regarding common values and goals, develop effective institutions that are able to mediate differences, agree to seek the 'common good,' and also share mutually agreed upon symbols and language" (2015, 317).

However, the basis for connecting the diaspora Jews of the world in a single state followed several different imaginations of citizenship, varying across diverse varieties of modern Jewish political thought associated with political movements, often categorized as political, labor, cultural, and religious Zionism. Some emphasized a unity among Jews that consisted of a spiritual tie. Others emphasized a togetherness consequent to shared persecution or a shared history as an exiled ancient diaspora nation. Some asserted a "natural" ethnoracial cohort. The materiality or immateriality of Jewish ethnicity remains contentious, particularly regarding the role of biological inheritance in guaranteeing Jewish identity. After these varying ideas as to what constitutes the Jewish nation, biological measures of Jewishness are becoming an increasingly important part of the Israeli national discourse. In contemporary Israel, Jewish ethnicity is often imagined as something rooted in the body, transmitted by genes, and shared by the world Jewry. A genetic understanding of Jewishness, however, represents a new way of imagining ethnicity in the young Middle Eastern ethnonation, with the roots of this biologization of Jewish identity lying in Europe.

ZIONISM AND JEWISH IDENTITY

The Zionist movement emerged in Europe at the turn of the twentieth century as a nationalistic solution to the so-called Jewish question on modern political terms. Different groups of Zionists espoused conflicting ideas about their explicit political goals and their religious sensibilities. So-called labor Zionists, influenced by Marxist-inspired reform in Russia, advocated a secular state and emphasized vigorous physical labor and the rejuvenating effects of working the land. Religious Zionism, on the other hand, emphasized a more diffuse spiritual unity as the essential condition that would make possible an ideal Jewish state. Different scholars have offered different explanations for what unites Jews, wherein the very "nature"—that is to say the core fundamental definition—of Jewish ethnicity and citizenship is "co-produced" (Jasanoff 2004) with the political telos of community-building pursued. In other words, the various dominant images of Jewish ethnicity and their performances must be considered in relation to their particular social, cultural, political, and historical contexts.

In the political philosophies of early Zionist thinkers, you can see the beginnings of concepts of Jewish citizenship and ethnicity that would eventually frame the establishment of the State of Israel. These Zionist thinkers conceived of diaspora Judaism—and by extension, the Israeli citizen or the "New Hebrew"—as a work of self-fashioning that would be possible when Jews were physically and/or spiritually relocated proximal to the epicenter of Jerusalem.

Austrian journalist Theodor Herzl was one of the key founders of political Zionism. His ideas had their roots in the ambivalent neo-Romanticism of fin-de-siècle Europe, that is, "between the fears and despairs of the post-Enlightenment *Kultur* and the respect and awe of post-industrialist scientific rationality, or *Zivilisation*" (Falk 1998, 590; emphasis in original). Herzl (1896) believed that attempts to assimilate Jews into European society were in vain since it was always the majority of each country who could decide who was a native and who an alien. He resented the idea of "belonging" as a criterion of privilege determined by national elites. He thought anti-Semitism to be a problem that would need to be solved by both global

Jewry and non-Jews acting in concert, thus transforming the "Jewish question" into a distinctly international political problem to be negotiated and resolved between nation-states on the world stage. In this regard, political Zionism's birth and strategic vision represents a reaction to the rise of anti-Semitism, European nationalism, and modern mythologies of ethnic purity, but importantly, Zionism is not an internal movement inherent to, intrinsic to, or a "natural" aspect of the Jewish diaspora in any unequivocal sense.

One of the trends in Zionist thought that sought to move against this kind of reflexive responsiveness to external political pressure and persecution was to root the Zionist movement on the organic plane of bodily labor, to take charge of the historical process by which the diaspora Jew would become the new Hebrew. Labor Zionism sought to reconcile Jewish history through a powerful ideology of Jewish nationalism and a strong desire to work hard and cultivate a robust Hebrew body. This ideology would demand an overhaul of Jewish political life and a transformation in diasporic traditions to inculcate the practice of Jewish nationalism at the level of the body, particularly through arduous labor practices.

The early labor Zionist and Ukrainian journalist Micha Josef Berdichevski underscores this imperative for historical rupture with diaspora Judaism, echoing Nietzsche's philosophical treatise on the "will to power":

> It is not reforms but transvaluations that we need—fundamental transvaluations in the whole course of our life, in our thoughts, in our very souls. Jewish scholarship and religion are not the basic values—every man may be as much or as little devoted to them as he wills. But the people of Israel come before them—"Israel precedes the Torah." (Berdichevski 1997, 294)

Accordingly, the Russian Zionist thinker Aaron David Gordon took up this thread to provide a theory of Jewish labor that he claimed would propel the Zionist movement forward to practical success. In the belief that Jews could become whole again by living the life of nature, Gordon likewise identified arduous bodily labor as the essential habit that Jews lacked:

> Labor is not only the force which binds man to the soil and by which possession of the soil is acquired; it is also the basic energy for the creation of a national culture. This is what we do not have—but we are not aware of missing it. We

are a people without a country, without a living national language, without a living culture. (Gordon 1997a, 373)

In Gordon's prognostications, a culture of labor would serve as the very glue or the "basic energy" that could tie Jews to each other, to the land, and, through that dialectical process, fill a deep lack and create a national culture to be enjoyed and sustained collectively. Further, for Gordon, "culture" was the dynamic and self-reinventing language of identity, and the new Hebrew Zionist movement would spread and be reproduced through joint labor, a manifest practice of nation-building. He painted a vivid picture of the "nature" of the labor Zionists' mutual solidarity with an acoustic metaphor:

> The ethnic self . . . is like choral singing, in which each individual voice has its own value, but in which the total effect depends on the combination of the relative merit of each individual singer, and in which each individual singer is enhanced by his ability to sing with the rest of the choir. (Gordon 1997b, 380)

While labor and political Zionists generally saw the move toward self-determination as a process of manifest vindication, the culturally inflected school of Zionist thought was apprehensive about this headfirst dive into a new Jewish culture. In fact, cultural Zionist Ahad Ha'Am rejected the Nietzschean will to power that Gordon backed so confidently, believing that hasty state-building and cultural refashioning would be a naïve mistake. He feared that Jews would no longer value the "moral good" and would instead elevate themselves above the general level of mankind. He doubted whether the moral development in the cultivation of a "Superman" ideal would serve the Jewish tradition well. He warned about potential regression: "Seeing that the goal is the mere existence of the Superman, and not his effect on the world, we have no criterion by which to distinguish those human qualities of which the development marks the progress of the type, from those which are signs of backwardness and retrogression" (1898, 225).

For Ahad Ha'Am, the Hebrew Superman is bereft of any moral compass to offer guidance toward an ethical future, with no agenda except the acquisition of power and instrumental domination of the immediate political environment. According to Ahad Ha'Am, Israel was already chosen by God for "moral development" (1898, 229); Israel has a moral purpose that is divinely inspired, and, as such, a transvaluation of its existing values would be an affront to God's will, disrespecting history and its "universal historical laws" (241). As to how to realize the ideal endpoint, Ahad Ha'Am urged Jews to reconcile the dualism of flesh and spirit—material and immaterial aspects of the Jewish individual—in a manner compatible with Jewish history and religious traditions, asserting that "the two elements in man, the physical and the spiritual, can and must live in perfect accord" (Ahad Ha'Am 1904, 150). In this regard, the historical dialectic is closed, and the Jewish spirit can be realized in concrete terms only through the establishment of the ethical Jewish state, and the state can be enlivened only with the healthy spirit of the committed and ethical citizen.

This formulation of the Zionist telos breaks with the labor Zionists' viewpoint in that it refuses to abandon Jewish religious tradition. More importantly, it sees the state as the materialization of spirit, which is to say that the dualism of spirit and flesh is folded into an ethic of state-building. In terms of realizing the birth of the state of Israel in practical terms, Ahad Ha'Am warns against looking forward with eager aspirations to modern novelty. Instead, Jews should look to the past for inspiration. Rather than tearing the fabric of Jewish traditions asunder, his conservative Zionist vision demands that the national ego emerge organically from history and law or, precisely, from the "foundations of the past" (Ahad Ha'Am 1904, 89).

Not all thinkers shared this conservative view regarding tradition. In profound opposition to Ahad Ha'Am's thoughts on preserving the foundations of Jewish history as though they were the inherited treasures of time, the Boyarin brothers (Boyarin and Boyarin 1993) praise diaspora Judaism's bricolage culture as a testament to the resilience and adaptability of Jews in the face of uncertain conditions. They pin Jewishness as precisely the ability to adapt, go unnoticed, and succeed as a "cultural trickster." They embrace the emergent cultural form of a dynamic diaspora Judaism. They reject the idea of Judaism as a fixed and essential cultural form: "Diasporic cultural identity teaches us that cultures are not preserved by being protected from 'mixing' but probably can only continue to exist as a product of such mixing. Cultures, as well as identities, are constantly being remade" (Boyarin and Boyarin 1993, 721). Though this kind of flux may be true of all cultures, they assert that diasporic Jewish culture makes it impossible to see "Jewish culture as a self-enclosed, bounded phenomenon" (721). This diasporic relational ontology of Jewish ethnicity, as defined by cosmopolitan experience, is fundamentally incompatible with a Zionist project of Jewish nationalism that sees the spatial sequestration of Jewish citizens in an exclusively Jewish ethnic homeland.

In distinction to such a fluid, contingent, and contextual conception of Jewish identity, religious Zionists typically emphasized the immaterial spiritual component of Jewish identity and the importance of gathering Jews in the land of Israel. Abraham Isaac Kook, the first Ashkenazi rabbi of British Mandate Palestine and an enigmatic and mystical philosopher of Judaism, exemplifies religious Zionism. Kook thought of Israel as "not something apart from the soul of the Jewish people" but "part of the very essence of our nationhood . . . bound to its very life and inner being" (Kook 1997, 419). This relation between soul and land that he professes cannot simply be explained away in political rhetoric or philosophy. Rather, he says, "human reason, even in its most sublime, cannot begin to understand the unique holiness dormant within our people" (419).

Writing outside of a rationalist "modern" discourse, or a dialectical tradition attempting to reconcile contradictions, Kook's mysticism transcends the realm of concrete politics and moves into the diffuse realm of the experiential Holy. "Deep in the heart of every Jew," he writes, "in its purest and holiest recesses, there blazes the fire of Israel" (1997, 421). Seeing Israel as an extension of the redemptive process that commenced with the exodus from Egypt, the "light" of Israel can be understood in his thoughts as being on the plane of a cosmic totality, being the final Jewish redemption with which history has been forever pregnant. Such messianic religious Zionism is far removed from the pragmatic action advocated by political and labor Zionists, but like cultural Zionism, it foregrounds the immaterial dimension of diaspora Judaism and the spiritual component of Jewish ethnicity. Religious Zionism does not, however, regard Jews as a race in a biological register.

In the late 1800s and early 1900s, however, before the establishment of the State of Israel, and in the post-Enlightenment milieu of secularization, Jews became understood as a racial category. Berman writes, "Jews themselves had helped construct racial typologies that classified Jewishness as a biological variant. Indeed, race language was a useful way to talk about Jewishness: it demanded little in the way of specific practice from Jews, and it seemed to guarantee Jewish survival as long as Jews continued to reproduce themselves" (2009, 16). However, racial constructions of identity also served hierarchical notions of racial superiority. According to Berman, "Race assumptions marked human difference in powerful ways, but they were also often employed to naturalize hierarchies among social groups" (16). Racial ideas also set Jews apart as fundamentally and unchangeably different from their Christian neighbors (Kaye/Kantrowitz 2007, 13) and often encouraged anti-Semitism.

After World War II, Jews turned away from biological understandings of Jewish ethnicity. Scholar of Judaism Jonathan Sarna attributes this evolution of thinking, "in response to Hitler, and in line with the teachings of anthropologists, they may have looked to culture rather than biology to explain the origin of ethnic differences" (2011, 108). On shifting away from a racial understanding of Judaism, thinker Mordecai Kaplan argued, "Jews should be understood as a 'distinct societal entity.' . . . What made a Jew a Jew was not what he or she believed, but how he or she lived. Religion, in other words, was a social phenomenon, and Jewishness, larger than religion alone, was a composite of social phenomen[a]" (qtd. in Berman 2009, 4).

The establishment of the State of Israel problematizes a single precise definition of Jewishness, since the state was founded on secular socialist principles, relies on *halakha* (religious Jewish law), and was built by waves of culturally diverse Jewish immigrants from Europe, North Africa, and the Middle East, all with varying levels of Jewish religious practice (Nesis 1970, 59). Maintaining a steady stream of Jewish immigrants has been a crucial facet of Israeli state-building, facilitating the integration of world Jewry, and fulfilling the state's mission as homeland and refuge for all Jews.[4]

The "authenticity" of Jewish immigrants who wish to participate in Israeli state-building "has been judged (often simultaneously) in both religious and bioethnic terms" (Burton 2015, 82). For example, the Population Registry Law 5725 of 1965 requires residents to enter both their *le'oum* (nationality or ethnic group)[5] and religion when registering for an identity number.[6] A 2013 Israeli Supreme Court case affirmed an earlier precedent and distinguished *le'oum* or nationality from secular citizenship; the court rejected the petitioners' request to list "Israeli" under the nationality rubric on their identity documents, which would reflect their citizenship and belonging to an imagined Israeli nation, rather than "Jewish," which reflects an ethnoreligious affiliation.[7]

So, it is clear that the various strands of Zionism that emerged in early twentieth-century Europe—labor, religious, cultural, and political movements that contributed to the establishment of the state of Israel—promoted distinct notions of Jewish citizenship. Political Zionism hinged on a relational ontology of Jewishness, with Herzl pointing to anti-Semitism as the intersubjective constitutive factor binding diaspora Jews with a common political goal. Labor Zionism emphasized "solidarity" and a shared culture of bodily practice, cultural Zionism emphasized the creative use of Hebrew and valued historical continuity, while religious Zionism has emphasized both a spiritual and material connection between Jews and the land of Israel. This disparate set of roots that yielded the Israeli state has grown from a heterogeneous entanglement of diverse political thought to yield a centralized state apparatus, with varying attitudes toward the social "nature" of Jewish citizenship as it is condensed into law and practice. In order to determine how these various layers of Zionist thought have led to the present case, in which Judaism can be attended to at the molecular level, as with "Jewish genes," we need to consider contemporary secular Israeli culture.

ISRAELI SOCIETY

In the early years of the Zionist movement, mainstream secular Zionist national identity became a hegemonic force. It can help to consider what Kimmerling (2005) calls "Israeliness." His analysis of Israeli society posits seven distinct "cultures" that constitute the pluralism of the country: the secular Ashkenazi upper class; the national religious; the traditionalist Mizrahim (Arab Jews, who have presumably always resided in the Near East, and North African Jews); the Orthodox religious; the Arabs; the Russian immigrants (especially since the fall of the Soviet Union); and the Ethiopians (who mainly immigrated to Israel in the 1980s and 1990s) (2005, 2). As of 2020, these groups form Israel's population of 9.1 million, of which approximately 6.7 million are Jews (Central Bureau of Statistics 2019). The Israeli state remains an immigrant settler polity that lacks a consensual social identity that unites these diverse groups, raising questions over its boundaries and positioning in the geopolitical environment of the Middle East. But despite the diversity and pluralism of the state's demography, a sense of a collective Israeli community has emerged (Kimmerling 2005). Kimmerling identifies the state, the education system, and the military as the three key institutions that help stabilize a sense of shared "Israeliness." But the Israeli state does not treat all of its citizens equally. Although secular Jews and a secular cultural life exist in Israel, it is not automatically clear whether "Israeliness" is a class of citizenship that necessarily requires Judaism at some fundamental level, necessarily excluding non-Jews from complete civic inclusion.

Israeli scholars Shafir and Peled (2002, 1) claim that Israel's principal moral political dilemma is thus the need to choose between the cardinal principles of the universalist commitment to being a Western-style democracy versus the particularist commitment to being an exclusively Jewish state. They argue that it is not possible to separate Israeli democracy and Israeli citizenship from its settler-colonial beginnings. Nor is it possible to separate these settler-colonial origins from the state's continued journey (1), since Israeli ethnonationalism denies the possibility of cultural assimilation to non-Jews as the discourse on citizenship incorporates nonpolitical cultural elements as critical determinants of assimilation. For example, in July 2018 the Knesset (the Israeli parliament) passed a law that defines Israel as the nation-state of the Jewish people.[8] Those who oppose the law see it as a challenge to democratic values that undermines the rights of non-Jews. In this context we need to look at how citizens are legally made, through the law that governs Jewish immigration (*aliya*).

JEWISH *ALIYAH* (IMMIGRATION)

Immigration of Jews in Israel is governed by Israel's Law of Return 5710-1950, which states, "Every Jew has the right to come to this country as an *oleh* [Jewish immigrant]."[9] The law is implemented by the minister of the interior.[10] "In conjunction with the Citizenship Law, which allows every *oleh* . . . to receive citizenship, it enables every Jew to become a citizen of the state, almost automatically" (Sapir 2006, 1239). *Oleh* is the noun for a Jewish immigrant to Israel and derives from the Hebrew verb "to rise, or ascend." The related gerund, *aliya*, meaning Jewish immigration, connotes the spiritual ascension imagined to take place with immigration to Israel. For the first twenty years that the law was in place, it did not define who was a Jew or provide guidance regarding who had the right to immigrate (Burton 2015, 79). In 1970, the law was amended to include a definition of Jew that reads, "For the purposes of this Law, 'Jew' means a person who was born of a Jewish mother or has become converted to Judaism and who is not a member of another religion."[11] The 1970 amendment also extended citizenship rights to family members of eligible Jews:

> The rights of a Jew under this Law . . . as well as the rights of an *oleh* under any other enactment, are also vested in a child and a grandchild of a Jew, the spouse of a Jew, the spouse of a child of a Jew and the spouse of a grandchild of a Jew, except for a person who has been a Jew and has voluntarily changed his religion.[12]

The amendment represented a compromise position between religious and secular perspectives (Altschul 2002, 1356). The amendment adopted the religious, *halakhic,* definition of a Jew—someone with a Jewish mother or someone who has converted to Judaism.[13] However, the amendment also extended citizenship rights to those who are referred to as "seed of Israel"—"a halakhic term that applies to anyone either born to a non-Jewish mother and a Jewish father, or having at least one Jewish grandparent" (Maltz 2015).[14] Thus, the law grants citizenship rights to those who are religiously Jewish but would not have Jewish biological links, such as Jews who have converted, as well as to those who do not have religious or biological connections to Jewishness, such as the spouses of Jews.

The 1970 amendment was a response to a controversial Israeli Supreme Court case that permitted the children of a Jewish father and non-Jewish mother to register as part of the Jewish *le'oum* or ethnic group in the Population Registry.[15] Additionally, "the amendment was intended to accommodate a small number of mixed nuclear families as the result of [this Supreme Court] ruling."[16] According to the Jewish Agency for Israel, "this addition not only ensured that families would not be broken apart, but also promised a safe haven in Israel for non-Jews subject to persecution because of their Jewish roots."[17] The amendment, therefore, expanded who was granted entry and citizenship but restricted who was classified as part of the Jewish *le'oum*, or nation.[18] So, if you are an Irish non-Jew, for example, and you marry an Irish Jew and you both move to Israel, you are entitled to Israeli citizenship but you would not be considered Jewish by religious authorities.

Since the law was amended, and especially since the fall of the Soviet Union in the 1990s, the population of immigrants in Israel has shifted significantly. Made possible by the changes in the Law of Return, as well as looser restrictions in the FSU that permitted residents to leave, Jews from the FSU have arrived in Israel en masse. According to some estimates, nearly a million people have come to Israel from the FSU under the Law of Return; at least a third of these are not Jewish according to religious law and by their own admission.[19] Many of these individuals had assimilated and secularized in the FSU, often abandoning Jewish religious practices and marrying non-Jewish Russians. Some statistics suggest that Russian immigrants have different feelings of Jewishness and belonging than their Israeli-born counterparts (Altschul 2002, 1360). Further, some non-Jewish Russians, who sought better economic opportunities, took advantage of the law and pursued entry through false documentation.[20] Although many Russian immigrants are Jewish by descent and are entitled to citizenship, their Jewishness is questioned by the Ministry of Interior, and they are often required to show additional proof. This proof has often been difficult to produce. Russian Jews, when marrying, have not signed Ketubot (wedding contracts) that for many immigrants can serve as evidence of family religious history.[21] These individuals face even more skepticism from rabbinic authorities, as many are not considered Jews under Orthodox Jewish law. According to Rabbi Hammer, "the position of the Jerusalem Bet Din of the Chief Rabbinate on these matters has been that regardless of the position of the [earlier religious teachings], they do not believe anyone coming from Russia without specific proof. Rather they must see a birth certificate and that of the person's mother."[22] Proof of a Jewish mother would be sufficient for religious authorities, and consequently the state, to recognize Jewishness and grant citizenship on that basis.

SECULAR AND RELIGIOUS JEWISHNESS IN ISRAEL

This context of suspicion, coupled with the discrepancies between eligibility for Israeli citizenship and religious classification as part of the Jewish nation, create many challenges for Israeli citizens who are not considered religiously Jewish. This is particularly difficult for those wanting to be in an interfaith marriage. Israel is governed by a dual legal system in which civil and religious courts have jurisdiction over various areas of the law. Based on the millet system adopted from the Ottomans, the laws governing personal statuses including marriage and divorce are under the exclusive jurisdiction of the religious courts.[23] Under this system, only Jews who are halakhically Jewish are eligible to marry in the religious courts, to belong to synagogues, or to be buried in Jewish cemeteries. There is no civil marriage in Israel (Burton 2015, 82). One of the main functions of the rabbinic courts is therefore to provide judicial rulings on whether a person is Jewish. For the many immigrants from the FSU, the rabbis follow a standard procedure that involves examining Soviet-era documents, such as birth certificates, that contain a citizen's nationality. There are good reasons to search for authentication of Jewish identity.

A large number of immigrants who are eligible to immigrate under the Law of Return are not religiously Jewish. One study, by demographer Sergio Della Pergola, suggested that, using the religious definition, there are roughly 14 million Jews around the world (people born to a Jewish mother), but more than 23 million people who are eligible for citizenship under the Israeli Law of Return (Nachshoni 2014). The Ashkenazi Chief Rabbi of Israel, Rabbi David Lau, knows of one family in which, "because of one Jewish grandfather who is buried in Moscow, over [seventy-three] people (his children and grandchildren) moved to Israel through the Law of Return" (Nachshoni 2014). This leaves a large segment of the population eligible for immigration and citizenship but ineligible to legally marry and have children as fully recognized members of the Jewish population.

According to a foreign ministry spokesman, the reported policy "to require DNA testing for Russian Jews is based on the recommendation of *Nativ*, an educational program under the auspices of the Prime Minister's Office to help Jews from the FSU immigrate to Israel" (Zeiger 2013). The Prime Minister's Office attempted to distinguish the purpose of the test as a secular immigration regulation rather than a marker of religious identity, and reported, "We're not talking about a test to determine Jewishness. We're talking about a test to determine a family bond that entitles [the child to] *aliyah*" (Silverstein 2013). By emphasizing the distinction, the Prime Minister's Office maintains the line between secular citizenship and religious belonging in the Jewish nation and thus reinforces a secular understanding of a biological kinship-based conception of Jewishness as opposed to a religious or practice-based view. Since biological imaginations of Jewish identity are becoming more common, we will now examine the case that spurred the state's announcement about using genetic tests for potential immigrants.

GENETIC BIRTHRIGHT

Nineteen-year-old Masha Yakerson, like many of her Jewish college-age peers, attempted to sign up for a Birthright Israel trip in the summer of 2013 (Zeiger 2013). A Birthright employee told Yakerson, whose family is from Saint Petersburg, Russia, that in order to prove that she was Jewish, and thus eligible for the trip, she would need to first take a DNA test. According to Birthright Israel's website,

> Taglit-Birthright Israel is a unique, historical partnership between the people of Israel through their government, local Jewish communities (North American Jewish Federations; Keren Hayesod; and The Jewish Agency for Israel), and leading Jewish philanthropists. Taglit-Birthright Israel provides a gift of peer group, educational trips to Israel for Jewish young adults ages 18 to 26. (Birthright Israel 2015)

The Birthright administrator claimed that the test was required by the Israeli consulate in Saint Petersburg and that a DNA test would be required if Yakerson wanted to make *aliyah* (immigrate to Israel). Yakerson's father called the policy "blatant racism toward Russian Jews" (Zeiger 2013).

In general, the requirements for teenagers from other countries to participate in Birthright are much less stringent than they are for Russians, and many participants do not meet strict definitions of Jewishness. For example, a similar post-college program, Masa, only requires that participants sign a document that declares they are Jewish, without any evidence to substantiate their claim (Maltz 2014). In fact, "since Taglit-Birthright doesn't accept candidates who have visited Israel before, its participants often come from non-affiliated homes, many of them the products of mixed marriages" (Maltz 2014). Historically, "trust was the default position" to determine whether someone was Jewish (Maltz 2014). If an individual claimed to be Jewish, he or she was believed. It is only more recently, in "an era of intermarriage, denominational disputes and secularization" that "Jews have ceased agreeing on who belongs" and doubt and skepticism have become the norm (Maltz 2014).

After the news of this student's experience made headlines, the Israeli Prime Minister's Office confirmed that many Jews from the FSU are asked to provide DNA confirmation of their Jewish heritage in order to immigrate as Jews and become citizens under Israel's Law of Return (Zeiger 2013). According to one source, the consul's procedure, which was "approved by the legal department of the Interior Ministry[,] states that a Russian-speaking child born out-of-wedlock is eligible to receive an Israeli immigration visa if the birth was registered before the child turned [three]. Otherwise a DNA test to prove Jewish parentage is necessary" (Zeiger 2013). This issue arose in Yakerson's case because her family was in the United States when she was young and her parents did not register her birth until she was three years old.

The State of Israel defines itself as the homeland of the Jewish people, making it ethnonational in its own self-image, with a particular theological commitment.[24] But this characterization does not sufficiently define the "legal nature" of citizenship in Israel.[25] It is not yet clear how a novel biological definition of Jewishness would impinge on Israeli law and basic rights to citizenship (See Abu El-Haj 2012; Goldstein 2009; Kahn 2005, 2010; Ostrer 2001). Moreover, as Israel has no written constitution, it is particularly important to consider the policy implications of novel biological determinations of Jewishness.

According to STS scholar Sheila Jasanoff, "periods of significant change in the life sciences and technologies should be seen as constitutional, or more precisely, *bio*-constitutional in their consequences" (2011, 3; emphasis in original). She elaborates that "revolutions in our understanding of what life is burrow so deep into the foundation of our social and political structures that they necessitate, in effect, a rethinking of law at a constitutional level" (2011, 3). However, the State of Israel has no formally written constitution:

> From its inception, Israel has never had a formal constitution, but only the Basic Laws. In its first years of existence, the government felt that it would be premature to set down in a definitive and binding way the nature and goals of the states and the Law of Return does not fall under the seven Basic Laws of Israel. Nevertheless, most believe that the Law would be given a distinguished place in a future constitution because the Law captures the ideology upon which the state of Israel was founded.[26]

Consequently, the recent discussions of genetic tests for Jewishness necessitate a rethinking of the specific Israeli law regarding the state's definition of Jewishness and, concomitantly, rights to citizenship. Following the controversial Yakerson case, Amnon Rubinstein, an author and professor at the Interdisciplinary Center Herzliya and former education minister and an Israel Prize laureate in law, wrote:

> In Israel, there are no DNA tests without court approval. These tests are only conducted when no other evidence of lineage can be found. In my opinion, when it comes to immigration to Israel, a mother's declaration regarding the identity of the Jewish father is sufficient—and there is no need for further proof. . . . There is no genetic test that proves conclusively whether someone is Jewish or not. There are certain tests for the genetic continuity of Kohanim (the Jewish priestly bloodline) and of various Jewish communities, and these prove the exceptional similarity between Jews and Palestinian Arabs. (Rubinstein 2016)

At this juncture, a closer look at the specific applications of Jewish genetics will prove instructive.

"JEWISH GENETICS"

Jews have been objects of racial classification and discrimination, but they have also applied racial concepts to themselves in various ways and for specific purposes (Bloom 2007; Efron 1994; Falk 1998; Goldstein 2006; Hart 1999, 2000, 2011; Morris-Reich 2006). In the last decades of the nineteenth century, European Jews were subjected to radical "biologization," particularly in Germany. There, Jews were presented as an Oriental race with distinct physical and mental qualities (Hess 2002). German anthropologists regarded Jews as a pure race formed by their practice of endogamy (Efron 1994, 20).

In some contexts, "race" was used to establish Jewish unity from within the Jewish community itself and was used to establish diversity and hierarchy among Jews. This was the case with Zionist literature that circulated in Mandate-era Palestine. Consequently, Hirsch argues that an Israeli formation of ethnic Jewishness owes its history to "the encounter of European Zionists with Eastern Jews, and from the tension between the projects of nation-building and of Westernization in the context of Zionist settlement in the East" (2009, 593). Hirsch observes that notions of "degeneration" and racial-eugenic "improvement" that migrated between the discursive fields of Europe and British Palestine helped to blur the distinctions among the biological, political, and social dimensions of Jewishness, making it difficult to separate the metaphor of eugenics from an emancipatory project of improvement via nation-building (2009, 596). In brief, Israeli Jews' imagination of a unified Jewish race has its roots in European diaspora host nations, twentieth-century biology, and essentialist nationalist imaginaries.[27]

Addressing the ways in which Jewish race science has transformed, and reemerged, in the twenty-first century, anthropologist of medicine Susan Kahn has identified three key ways in which Jewishness has now entered the molecular realm, with genes being defined as Jewish in three major ways: population genetics, genetic testing for both disease and Jewish identity, and human ova and sperm donation in the domain of assisted conception (2010, 21). In these different conceptual arenas, "Jewish genes" and Jewish inheritance are determined in markedly different ways.

In relation to population genetics, or "tracing Jewish history through DNA," Kahn claims genetic studies must be situated within the larger sociopolitical context, wherein the meaning of claiming Jewish identity can make a direct impact in terms of access to rights and resources (2005, 181). As reviewed above, Israel's Law of Return, the state's commitment to helping Jews come to live in Israel, makes it important to have verifiable evidence of "authentic Jewishness." But marginalized Jewish communities already in Israel may benefit from proof of "authentic Jewishness." The marginal groups of the Beta Israel of Ethiopia, the Kuki-Chin-Mizo from Northeast India,[28] the Bene Ephraim from India (Egorova and Perwez 2010, 2012), or the Lemba people of southern Africa, for example, could perhaps benefit from genetic evidence to support their claims to rights and equality. The Lemba people not only claim descent from a tribe of Israel with descent passed from father to son, and maintain some Jewish traditions such as a kosher diet, but Lemba men also possess a "Jewish genetic marker," the Cohan Modal Haplotype (CMH)—a genetic signature that has been identified among Sephardic priests in the Jewish population—with a frequency similar to that in the general Jewish population (in just under one out of every ten men).[29] This adds support to their demands to be regarded as equals to the traditional elites. But as Kahn reports, Jewishness, as determined by genomic analysis, is embodied as "statistical probabilities that DNA haplotypes will be more prevalent" within groups, and cannot say with certainty whether an individual is Jewish or not (2010, 21).

The CMH, the "Jewish DNA haplotype" that has received the most attention, was first publicized in the scientific journal *Nature* (Skorecki et al. 1997), in a study that identified six differences in the DNA sequence of male Jews that self-identified as Cohens. A haplotype is simply a group of alleles that are inherited together and consequently can be used to measure relatedness among individuals. It was thought that the "Cohanim" signature represents the inheritance of more than 100 generations from the founder of the patrilineal genetic line, with the signature traced to a date more than 3,000 years ago, in accordance with the oral tradition that the Cohens (Jewish priests) maintain a line of patrilineal descent from Aaron, the first Jewish priest (Kahn 2010, 14). In line with the cultural tradition of patrilineality, the CMH is found only on the male Y chromosome. However, since the Y chromosome contains mostly noncoding DNA, sequences that are not thought to translate into a physically expressed trait, it is unclear whether identification of the Cohanim signature holds any valid indexicality as to the nature of the bearer's body in terms of a physiological or biometric characteristic, even though it might be read as a valid inscription of ethnic history.

This sort of ambiguous phenotypic implication is not the case with inheritable diseases, however, for which DNA mutations carry a higher likelihood of developing a real disease. Indeed, European Jews are generally susceptible to a range of inherited diseases that are associated with identifiable genes. Common inheritable diseases among European Jews are Tay-Sachs disease, Canavan disease, Gaucher disease, familial dysautonomia, Niemann-Pick diseases, and Huntington's disease, making it important that bearers of the causative gene do not pass the disease to their children (Dor Yeshorim 2015). Consequently, there have been moves to test individuals for genetic markers of disease, either before they form partnerships or before they choose to have children together. The Brooklyn-based organization Dor Yeshorim, for example, established a database of DNA comprising samples from young Ultraorthodox Jews in high school (Kahn 2005, 181). The samples are cross-checked so that genetically incompatible matches between prospective marriage partners can be recognized in an effort to reduce the occurrence of genetic diseases in the community.

The Orthodox community has generally embraced the available genetic tests, but a concern remains in the community about "dangerous eugenic overtones" (Kahn 2010, 17). That said, it remains unclear whether the use of genetic tests for diseases common among Jews is contributing to a reductionist rationality that a Jewish disease is evidence of a Jewish body, or indeed the existence of a Jewish biological race. In relation to ongoing research on diseases in the Ashkenazi Jewish population, however, Mozersky and Joseph argue that ethnic genetic medicine "reiterates a shared history and addresses culturally salient issues" (2010, 425), which in turn both "encourages active participation" and "contributes to a particular version of population" (434). This finding accords with the ethnographic study of medical genetics by Fujimura and Rajagopalan that "analyzed how scientists produce simultaneously different kinds of populations and population differences, sometimes by appealing to popular categories of race, ethnicity, or nationality, and sometimes to 'genetic ancestry'" (2011, 22). They conclude "that the invention of new genetic concepts of ancestry relies on old discourses, but also incorporates new knowledges, technologies, infrastructures, and political and scientific commitments" (22). Genetic evidence is thus lent meaning in the historical context of its interpretation, with all of the beliefs and commitments that shape the identities at play.

In the context of Jewish assisted conception, it should be noted that there is a strong association between fruitful reproduction and Jewish tradition. The Orthodox community has consequently been receptive to the use of technologies to assist with fertility, and many rabbis permit the use of genetic donor material to circumvent a range of adulterous, or incestuous, unions (Kahn 2005, 184). Moreover, since Jewishness is traditionally passed from mother to child, non-Jewish sperm can also father a Jewish child if the mother is Jewish. However, the inheritance of Jewishness may be problematized if a surrogate mother carries a baby.

The question is whether a baby who has genetically Jewish parents, who donate the egg and sperm, but who is carried to gestation by a non-Jewish surrogate, will be Jewish. A case of this resulted in a rabbi from New York opining that the baby technically had three parents, and because the surrogate was not Jewish, the child was not Jewish (Chesler 2013). Believing the problem more complex than deterministic genetics or notions of modern biology, he reasoned that if motherhood involves both giving a child DNA and giving birth, and if science can now bifurcate these roles, then we have the condition of having two mothers. For a child to be Jewish, both mothers must then be Jews.

THE RELATIONSHIP BETWEEN SCIENCE AND SOCIETY

Despite the ambiguity of Jewish genes, genetics is becoming a way of imagining the limits of the Jewish population. Although genetic legitimation might be meaningful only if rabbis or others in power recognize it as a verifiable source of knowledge, there is a potential for Jewish genetics to be used to manage populations through "biopolitics," the governance of life itself (Foucault 1977, 2010; Rose 2007). Barbara Prainsack (2006) has argued that Israel's permissive laws regarding the use of artificial reproductive technologies can be traced to their utility in tackling Israel's "demographic problem," that is, in maintaining a Jewish majority. Moreover, she finds that Israel's pro-natalist culture rests on a notion of "risk" to the population that serves to bolster the state's mandate to reproduce the nation at the level of individuals. Prainsack writes, "The 'demographic threat' that the Jewish majority population in Israel will be outnumbered by non-Jews in the not too distant future provides a context of risk to the discourse on 'Israeli cells'" (2006, 173). In this context, genetics offers a way to imagine instrumental control over the demography of the state.

In the admittedly unlikely eventuality that genetic tests are routinely mobilized with efficiency to determine rights to citizenship in Israel, the foregrounding of Jewish genes as the basis of adjudicating cases of Israeli citizenship would nonetheless be a novel form of governmentality. We would be seeing the management of a population by a state through ethnic genetics. In facing the potentiality of genotyping citizenship, it is necessary to read this potential future development as the state using secular technoscience in an attempt to achieve a stable future.

Such a development, however, would be contested, particularly by religious Jews. When it comes to genetics as a means of testing Jewishness, many rabbis remain skeptical. One rabbi said he believed genetics could be a "consultant" to *halakha*, Jewish law (Wheelwright 2013). However, he worried about the newness of the technology, as well as the "binary yes or no of DNA analysis," which is inconsistent with the "cloudiness and argumentation [that] is built into the theocratic polity of Israel" (Wheelwright 2013). In this view, ambiguities should be resolved by debate rather than by genetic tests. For other rabbis, concerns remain about the "dangerous eugenic overtones" (Kahn 2010, 17).

Nonetheless, genetic tests offer the possibility of legitimizing those whose Jewishness is often questioned. In one case, an Eastern European woman had lived in Israel for twelve years and sought rabbinic permission to marry. She "had documents affirming that her paternal grandfather was Jewish, but no proof of Jewishness on her mother's side save her own testimony. To bolster her claim for a marriage license, the woman went to a commercial gene-testing service and had her DNA analyzed, specifically her mitochondrial DNA" (Wheelwright 2013). The DNA test "tipped the balance in her favor" and the "rabbi granted her a marriage license as a bona fide Jew." A genetic definition of Jewishness, however, breaks with the traditional *halakhic* law and reconfigures the terms of authentic belonging recognized by the Jewish state.

But genetics is by no means important for all Jews to authenticate their sense of belonging. In an article in the *Jerusalem Post* titled "Should Jewishness Be Determined by a Genetic Test?" the author interviewed recent *olim*:

> In 2011, Boris (pseudonym) found out he was Jewish after his grandmother on his mother's side told him on her deathbed that she was a Jew. She had grown up in a small village in Ukraine and as a teenager was sent to Auschwitz after the Nazis invaded. "Her entire family was murdered—parents and siblings—and after surviving the war she moved back to Ukraine and made a promise to herself that she would forget her past and Jewish roots. She married my grandfather, a native Ukrainian, a few years later. She never told my mother that she was Jewish. I get the feeling my grandfather knew, but they brought her up as an agnostic. My parents brought me up as agnostic as well, but I always felt there was something more." Boris, an only child, says he was surprised but not shocked by the revelation. "A year later, I went on a Birthright trip to Israel and after the visit I knew I wanted to live here. I finished my university studies in Ukraine and came here. I know that I'm Jewish even though I have no documents to prove it. I can feel it and no genetic test will tell me otherwise. If the time comes when I have to take a test, I won't because I know I'm Jewish." (Chernick 2017)

For those like Boris, Jewishness is grounded in personal and familial biographical experience and does not need to be authenticated by an objective science. For him, the truth of a dying grandmother could not be overturned by a genetic test.

In Masha Yakerson's case, however, genetic testing was used as a barrier to prevent access for someone who meets the expansive definition laid out in the Law of Return, but still was not "Jewish enough." For the Yakerson family, the turn to genetics has had strange results. Although Masha was ultimately denied access to the ten-day Birthright trip to Israel, her older sister, Dina, reportedly immigrated to Israel as an *olah* in 1990 (Zeiger 2013). For a test intended to measure family bonds and verifiable Jewish heritage, in this case, the turn to genetics actually failed to provide a consistent or accurate measure of familial connections. Rather, it would seem that reliance on genetics might achieve little more than flexing of the muscles of state power, a performance of bureaucratic rationality.

Depending on how the state uses this technology, the Israeli government's potential use of genetic testing to determine eligibility for citizenship or other rights can be interpreted in several possible ways. It could be a sign of a more restrictive immigration policy that seeks to guard access to the rights and resources of the state. In this interpretation, and in light of the economic challenges faced by many immigrants, it could be an attempt to alleviate unemployment and reserve economic prospects for those already in the country. Similarly restrictive policies have been advanced that require Jewish verification from those seeking temporary student or work visas as well (Maltz 2014). These temporary visas do not even involve the full benefits associated with permanent immigration and citizenship and suggest that more is at stake than merely guarding resources. One rabbi, who has dedicated his life's work to helping potential immigrants navigate the rabbinic bureaucracy, explained: "What we are witnessing is the creation of a culture of xenophobia in the corridors of power in Israel. . . . It manifests itself in the way we treat people born Jewish who don't fit the description of what a Jew should look like" (Maltz 2014).

The tests may also become a means to expand the pool of potential new Jewish immigrants who have verifiable ancestral ties (Maltz 2015). For the Bnei Menashe community of northeast India, Jewish genetic tests could become a way to recognize different and broader articulations of Jewish identity and thereby expand the limits of who has legitimate connections to the Jewish community. The potential move to acknowledge, legally, genetic tests for Jewishness could equally shift some of the authority away from the rabbis, who currently hold much power over entrance to the

Jewish community, and toward scientists, who may be more open to recognizing objective and secular manifestations of Jewish identity.

The varying secular/religious rationalities at play in "Jewish genetics" point to the ambiguity or outright contradictions between the field of genetics and rabbinic law in determinations of Jewish ethnicity. On the one hand, geneticists make claims that ancestry can be determined on the basis of DNA sequences passed from father to son, even though non-Jewish sperm may be used to father Jewish babies. A baby without any Jewish DNA could, however, be a complete Jew. Indeed, the majority of contemporary Orthodox rabbis agree that a child conceived with an egg donated by a non-Jewish woman is considered Jewish as long as the fetus is gestated in a Jewish womb (Kahn 2005, 184).

In Orthodox discourse, Jewishness is not a genetic issue. In the rabbinic imagination, the identity of the birth mother determines Jewishness. A child conceived with a non-Jewish egg and a non-Jewish sperm would be considered fully Jewish once it is born of a Jewish womb. An interesting contradiction thus appears. Although Jewishness can be traced genealogically by reading DNA up the paternal line,[30] as is the case with the Cohan Modal Haplotype, Jewishness can be reproduced only in the present, that is "passed on" through the maternal line through the process of gestation in a Jewish womb.

The flexibility and the gendered dimensions of Jewish identity are highlighted by the ambiguity of "Jewish genes" in the transmission of Jewish identities through birth. It might therefore be more sensible to think about Jewish genetics as a discourse that mediates collective visions of peoplehood depending on what it achieves rather than on where it fails. The epistemic qualities of "Jewish genetics"—their validity and consistency—can be viewed as secondary to the event that is achieved in the political present. "Jewish genetics" is a technical iteration of identity politics and a genre of discourse that mythically reinforces the imagination of the singular nation. It cannot be meaningfully discussed without recourse to the specific moment within which the epistemic value of claims to genetic identity affords utility. In other words, we need to consider "Jewish genetics" within the context of relations of power between citizens and their government, as well as between those who are excluded from both citizenship and recognized Jewishness.

In public discourse "ethnic genetics" reifies the Jewish nation as a unified entity. Indeed, the epistemic value of "ethnic genetics" and the political milieu appear to be "co-produced"—they beget and stabilize each other (Jasanoff 2004). Simply put, without a Jewish state in the Levant, questions over "Jewish genes" would probably hold a very different kind of importance and interest. Crucially then, ethnic genes may serve to make states into more stable political realities, while states simultaneously create the conditions for the meaningful misrecognition of genetic material as bearing an essential identity. The potential for "Jewish genes" to serve as a measure of inclusion in Israel makes this patent, but regardless of what happens in Israel in the coming years, the imagination of "Jewish genes" has entered both Israeli public discourse and the state's political imaginary.

In facing the potentiality of genotyping citizenship and cataloging the "genomic citizen," it is necessary to recognize an attempt to imagine a future for the Israeli state through the visions mediated by a secular technoscience. This in itself is not novel, since secular visions of the Israeli state have previously been described in relation to science and technology, for example with David Ben-Gurion's "scientific utopianism" and his "million plan" to bring a million Jews to Palestine (Barell and Ohana 2014), or with the Israeli geneticists who in the 1950s applied their science to establish a national identity and confirm the Zionist narrative (Kirsh 2003). Kirsh finds that Israeli geneticists unconsciously internalized the Zionist narrative, and Zionist ideology is evident in their genetics research, which evidences their beliefs about the origins and history of the Jewish people. But the latest possibility of genetic tests being used to decide citizenship could transform the very definition of the Jewish political subject. At issue is the possibility of a novel form of governmentality in the distribution of citizenship. Regardless of the validity of genetic tests for Jewishness, this possibility itself entails a unique iteration of Jewish political thought, a geneticized articulation of a secular Zionism that foregrounds the subject's genetic code in the imagination of civic inclusion.

One of the key institutions that have made this discourse of Jewish genetics possible is the National Laboratory for the Genetics of Israeli Populations. As described in the next chapter, I went there as an ethnographer, hoping to learn how this science of identity is produced. How, I wanted to know, does Jewishness become a category of analysis in genetics research, and how does this research foster an imagination and discourse of a genetic collectivity?

3 THE ISRAELI BIOBANK: A NATIONAL PROJECT?

THE NATIONAL LABORATORY?

"It must be a mistake," says David Gurwitz, head of the Israeli biobank, the National Laboratory for the Genetics of Israeli Populations (NLGIP) at the Sackler School of Medicine at Tel Aviv University. I had told David of the recent news article reporting that Prime Minister Benjamin Netanyahu had announced that genetic tests might be used to determine whether immigrants were Jewish. Judaism, David told me emphatically, is a religion, and cannot be determined by a genetic test. If only it were so simple, I thought. I knew by then that popular discourses of genomic citizenship were too weighty to be abolished by something as inconvenient as scientific invalidity. The discourse of genomic citizenship exceeds the science; the "bio-nation" is discursive all the way down. The rhetoric of Jewish genetics is far more about identity than about genetics.

As an ethnographer, I was disappointed by my experience at the NLGIP. In the context of the wide circulation of gene talk and the potential biopolitical role of genetics in Israeli society, I had expected the NLGIP to be replete with research and discourse concerning the genetics of Jews. I was expecting to find work on the genetic nature of the Jewish nation and perhaps also on the genetic basis of a return to Zion. But these expectations were not met. It turned out that the NLGIP is tightly woven into the fabric of Israel's burgeoning secular technoscience. It is concerned with an unmarked global science and the imagined move toward a future era of precision medicine. The Zionist pioneer at the NLGIP is, rather than a religious-nationalist fanatic, the secular humanist scientist pushing the boundaries of global biomedical progress forward. This is the Zionism of twenty-first-century secular global modernity—in Tel Aviv, global scientific hotbed.

In 2015, within a month of arriving in Tel Aviv for a year of fieldwork, I had sent an email to David to request a meeting. It was a hot and humid August, and each morning I would ride my high-speed electric bicycle to the Tel Aviv University campus where I was attending a one-month Hebrew language course. David responded to my message within a day, saying that he would likely be of little help to me since I should probably be speaking to genetic counselors, but that he would be glad to meet with me nonetheless. I had harbored expectations that this "national laboratory" would be the perfect site for apprehending the collision of national imaginaries and basic science, so I disregarded his discouragement. I presumed he didn't see the kind of anthropological connections I was pursuing. We scheduled a meeting and the following week, and after navigating the labyrinthine corridors of Sackler, I found his office, tucked in the far corner of the building's seventh floor, beside his small wet lab and adjacent to the national biobank's storage room.

I introduced myself and explained a little about my background in biology and anthropology. As I had already had three years of Hebrew classes and was fairly fluent in the language, I initiated the conversation in Hebrew, but after he responded several times in English, I moved to English. On several occasions, I got the impression that English was the appropriate language for discussing science with outsiders. English, after all, is the lingua franca of global science. I told David I was interested in studying the biobank that the NLGIP manages from an anthropological perspective, with a focus on ethnicity and national identity. After a half hour of conversation about the research activities of David's lab, it was clear that I would be able to join the lab and "help out" with some experiments. I enthusiastically volunteered to offer my laboratory skills as a way in. More specifically, one of his graduate students, Keren Oved, was working on a neurobiology problem that I might be able to assist with, considering I had some expertise in the area.

I followed up with an email the following day, and at our next meeting, David introduced me to Noam Shomron, a younger, up-and-coming faculty member with a larger genomics lab in the same building. Noam had recently returned from a postdoc at MIT, and he was now establishing his own lab. Noam, David told me, also has an interest in the field of science and society, as well as that of bioethics in relation to genomics. I felt like I had landed on my feet, having secured access to the NLGIP as a visiting scientist and getting introduced to another important player in Israeli medical genetics. I surely had gained access to a rich site where the melding of Zionist ideology and molecular genetics would obtain in ethnographic richness. That was my hope and expectation.

About a week later I met with Noam in his lab, which was much larger than David's and generously decorated with oil paintings on canvas, mostly of animals and natural landscapes, which, I found out later, were all painted by Noam. I found Noam in his office, where he offered me a seat and asked me to explain a little about my interests. He listened attentively and patiently as I explained my research interest in the context of my biography, eventually getting to how I ended up studying Israeli society. I told him that I was interested in the relations between genetics and the cultural context and, in particular, the ways in which racial and ethnic identity play out in the management of genetic data. When I eventually paused for his response, he said conclusively, "I think you've come to the right place," before giving me an overview of the lab's work and suggesting some ways I could get involved.

Noam made me feel immediately welcome and told me about the lab's ambitions and overseas collaborations. He showed me a small biomedical device he had on his desk, only slightly bigger than a matchbox, which he told me he had "on loan" from the manufacturers. It was a prototype for the latest high-speed portable genomic sequencers, and his lab was trying it out. Noam's openness, confidence, and willingness to invite me into his lab impressed me, even flattered me. Crucially, I was convinced that his lab would be a place where I would be able to see the management of genetic data of identified ethnic populations.

There was also an obvious way for me to integrate into the lab. David had already paired me with Keren, a PhD student working with both Noam and David, and with whom I would work on a project investigating the molecular basis of resistance to antidepressant therapy (specifically, selective serotonin reuptake inhibitors, or SSRIs) and Noam offered me bench space next to

Keren. I sat at a small desk, between Keren and another PhD student. I began attending the lab in the morning, sitting and working on my laptop, listening to the conversation in the lab, and asking lab members what they were doing or to explain a little about their projects. The conversations in the lab were usually in Hebrew, and I participated and conversed with the lab members in Hebrew (many of whom didn't speak English well), but Noam and David usually spoke to me in English. Though I had secured excellent access to the basic research practices of the lab, it was not clear how the Israeli biobank, and the diverse genetics research in Noam's lab, fitted with my ambitious theoretical questions about the molecularization of ethnicity and the discourse of Jews as a biological nation. I began to wonder whether I would find what I was looking for, that is, the imagination of the national collective shaping the handling of genetic data. Perhaps if I had had more experience doing ethnographic research, I would have been more aware of how much I did not know. I might have been more accepting that I was yet to learn what would be important, and I would have been less invested in any particular outcome.

BIOLOGICAL NATION?

Israel is a country of nine million people with a relatively equal balance of men and women and with a distribution across age groups that is typical for a growing society (see figure 3.1).

The Jewish population of Israel stands at about 6.7 million people (74.1% of the population), with the Arab population being about 1.9 million (21%). Also, about 448,000 (4.9%) people are either non-Arab Christians or are listed as "no religion" in the civil registry (Central Bureau of Statistics 2019). Unlike other Middle Eastern states (such as Qatar, the United Arab Emirates, Iraq, or Yemen), in Israel it is the demographic majority (Jews) that can be considered the hegemonic, ruling group. This fact is significant in considering the role of biobanks in providing a way of imagining the national cohort and the limits of belonging. The demography of the state's territory and indeed the disputed territories (under the administrative control of Israel with the support of the Palestinian National Authority) also need to be recognized in analyzing any representation of a

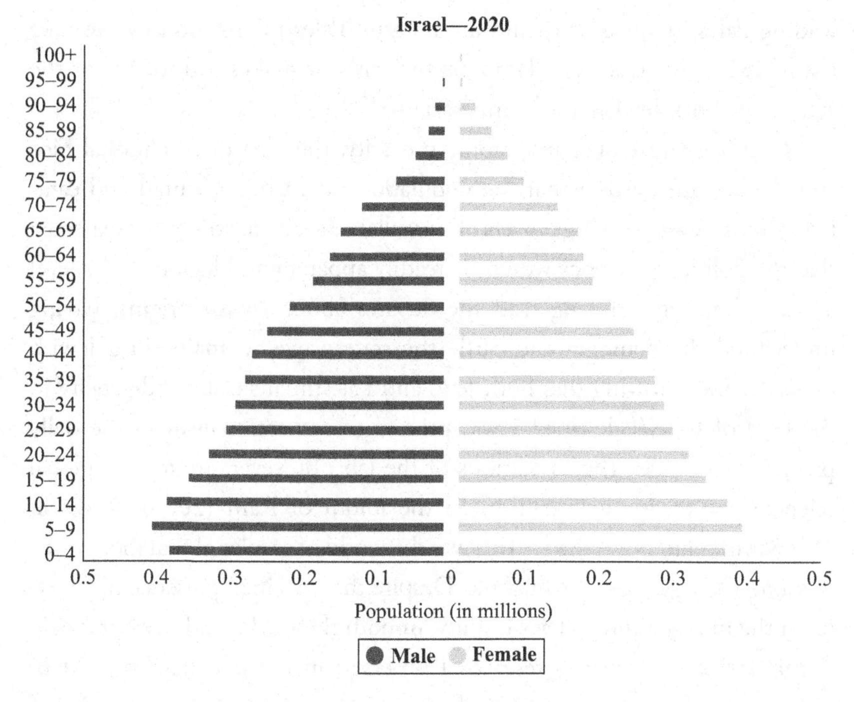

Figure 3.1

Israel population pyramid. A population pyramid illustrates the age and sex structure of a country's population and may provide insights about political and social stability, as well as economic development. The population is distributed along the horizontal axis, with males shown on the left and females on the right. The male and female populations are broken down into five-year age groups represented as horizontal bars along the vertical axis, with the youngest age groups at the bottom and the oldest at the top. The shape of the population pyramid gradually evolves over time based on fertility, mortality, and international migration trends (CIA World Factbook 2020a).

Jewish national population. Indeed, since the 1947 partition plan, Jews, "with 32 per cent of the population . . . were awarded 55 per cent of the land and 80 per cent of the coastline" (Anderson 2015, 34). In this arrangement "Arabs, with 68 per cent of the population, were allocated 45 per cent of the land" (35). Since then, Israel has expanded its sovereignty beyond this initial plan, and birth rates and rates of immigration have varied among the two populations: "Heavy Jewish immigration and high Palestinian birth-rates have ended in the rough parity at which they stand today—Jews leading Palestinians by a dwindling margin, Palestinians soon to overtake them" (35). This is Israel's demographic crisis. It makes it difficult for the state to be both Jewish and democratic.

In this context of demographic precarity, the idea of a "national biobank" made me suspect that the population might be measured and cataloged for the ever-growing powers of surveillance available to states. I expected that the politics of science would be readily apparent and legible at the level of the laboratory. Having read the debates about Jewish origins, having understood the demographic battle the state wages to maintain a Jewish majority, and knowing that both Jews and Palestinians claim indigeneity in the land of Israel/Palestine,[1] I was primed to see nationalism in the daily practice of science. The aspirations of the labs, however, are toward global science, not Jewish nationalism. In the idiom of Ram (2008), I was in "McWorld," and not "Jihad." This was the world of secular global modernity, not a mythology-soaked ethnocult. Despite the sustained presence of genetics in the imagination of Jewish unity, in both the media and discourse over immigration, the genetics research I observed in the labs fell far short of delivering epistemic grounding and evidentiary footing for the bio-nation.

This ethnographic reality became progressively apparent during my time at the NLGIP, all the while I had the sinking feeling that my fieldwork was falling flat. However, I continued to question whether this is a facility that furthers national biometrics or the surveillance of populations, whether it serves to manage or control the demographic nature of the population, whether it projects an image of the population in service of state-building. Is the biobank a biometric apparatus of the state, serving to survey the population? Lebovic characterized biometrics as "the archiving of biological data, based on the surveillance and control of bodily images in public space" (2015, 843), and adds, poetically,

> The physiognomy of our age has been secularized, automatized, digitally coded, visually metaphorized, privatized, and depoliticized. In other words, the wide spread of biometric systems proves that the modern aestheticization of politics, which lasted from the eighteenth century to the twentieth century, has turned into a system of hidden and fragmented biological control. (Lebovic 2015, 842)

Similarly, in relation to the biopolitics of the body, Comaroff and Comaroff assert that "in the mass-mediated *imaginaire*, science has come to be the panacea for the policing of everything, but despite its mythologizing in popular discourse, its methods, for all their utility, do not remove the doubts, deficits, and indeterminacies that beset enforcement everywhere" (2016, 47). Science, despite its imagined promise to know and render clear, often leaves doubts and raises more questions about certainty.

These contentions, of course, posit a bifurcation between the mythologizing popular uptake and circulation of scientific imaginaries, like the bionation, and the application of scientific methods in managing populations that has become known, following Foucault (2010), as "biopower." In other words, there exists an uncoupling of the circulatory semiotics of science as biopower from the instrumental capabilities its technologies achieve in practical terms, even while the material and the semiotic are two sides of the dialectic of the discourse of power and knowledge. If science is both mythology and instrumental control, a national biobank ought to be suspiciously interrogated as a direct or indirect effort to control populations.

In my fieldwork, I became attuned to the biopolitical potential of the apparatus. I considered whether the biobank could be read as a nation-building device for a technocratic and secular society. But it became more difficult to make the connection between the genetics research I observed and the potential downstream outcomes. As the question that I brought with me to the field was how the biobank fits with previous imaginations of bodily substance as a site of national instantiation, and since the ethnographic character of the laboratory life did not yield the expected rich discourse of nationalism, I began to investigate the genealogical origins of the biobank, its history, practices, and imagined purpose and utility. Ultimately, this decision expanded the frame of my research to consider the global context of biomedicine.

I begin to address broader issues in genetic medicine with a vignette from the work of one of Noam's master's degree students, Yaron, whose work investigates genetic markers for Parkinson's disease in Ashkenazi Jewish populations. The purpose of this case is to demonstrate the precise way in which ethnicity comes to both matter and not matter in the genetics research I observed in Tel Aviv.

"MEDICAL IMPLICATIONS FROM INVESTIGATION OF THE JEWISH EXOME"

One morning while I was sitting at my desk writing an op-ed for the journal *Genetics Research*, which Noam edits, Noam asked me if I'd like to join in a lab meeting. I grabbed a chair and joined the lab members in the computational "dry" bench area. I introduced myself to the group of students, some of whom I hadn't yet met, and then the master's degree student, Yaron Einhorn, began his presentation. Yaron was working in the area of bioinformatics, and his presentation was highly relevant to my research. He discussed the medical implications of an investigation of the "Jewish exome" (coding genes, with a possible biological function). The talk was in Hebrew, but the PowerPoint slides and graphs were in English. This, I later found out, was usual in scientific presentations in Israel. In comparison to my personal experience in academic communities in the United Kingdom and the United States, there was an air of casualness to the presentation, which I would later learn to be typical of Israeli life and which is characteristic at Tel Aviv University. Yaron was wearing flip-flops, cargo shorts, and a sleeveless T-shirt, with headphones sitting on his neck while he presented to the group. His talk was titled "Medical Implications from Investigation of the Jewish Exome." He provided a summary of the research he had completed for his master's thesis and explained how populations and ethnic groups are measured and identified within population genomics, as a way to identify biomarkers that could aid the understanding of Parkinson's disease.

The study was an analysis of the genomic data from a cohort of 74 Ashkenazi Jews between the ages of 39 and 85, 54 of whom had Parkinson's disease. The goal was to identify possible genetic markers that could lead to a better understanding of the genetics of the disease. If you can find genetic markers that are present in the Parkinson's patients that are absent in the healthy patients, then you may be able to learn more about which genes help cause the disease. The data for this research, however, did not come from the biobank. Since it is too expensive to sequence all the individuals represented in the biobank, and since there is poor medical information about the biobank participants (the biobank is primarily a tissue repository and not a health database), the data for this study came from an open-access database called the 1000 Genomes Project (from which it is possible to freely download the genomic data of individuals), and ExAC—a database of exons (coding DNA) of populations.

One of the first few slides of Yaron's talk dealt with sorting out the participants or seeing where the participants lie as a related genetic cohort. I was surprised to see that the individuals were broken into distinct racial groups at the beginning of the analysis. In analyzing the genetic variants that were present in the Ashkenazi population, Yaron compared their incidence to both European and African "reference populations." Neither Yaron nor the lab members questioned the boundaries, or identities, of the racial categories. Rather, the racial identities of these groups were used as a "reference" within which to sort out the newly identified genetic variants. I was interested in how and why groups can be easily formed around racial identity when the liberal consensus is that race is arbitrarily and socially constructed, that humans are vastly more genetically similar than they are different from one another, and that science that reinscribes the idea that racial groups can be segregated in biological terms is dangerous by virtue of its implications in naturalizing difference. I wanted to know why racial reference populations were necessary.

With genomic readings, I learned from the talk, it is not an essential, visible, characteristic that determines an individual's racial group. Nor is it self-identification. Rather, individuals are deemed related based on their shared genetic variants, and the likelihood that individual groups are associated, historically, is determined by the degree to which they share these differences. You could say, therefore, that having genetic variants similar to others in one's own racial group is the basis for comparisons across groups. Consequently, the natural clustering of variants leads to the labels from the geographic origins of these individuals. People thus get bunched into racial groups that have common ancestry.

However, these differences are not necessarily phenotypic; that is, they may not be visible, or physically expressed, differences. These variants may not have an important biological function at all, but may rather, simply, function as traces of history. By establishing racial groups as reference populations for these genetic studies, the "difference of difference" is compared, such that the high degree of genetic similarity between groups is somewhat occluded. But racial type or geographical background had no further importance or emphasis in Yaron's research. He was interested only in the role that variants may play in the development of Parkinson's. He was not investigating the legitimacy of categorizing groups based on racial categories. If, for example, he found a variant that could be traced to the development of a particular biological function that is important in the development of the disease, the race or ethnicity of the individual who has the variant would not matter. The important finding would be that the genetic marker could be used to predict the disease in any individual, regardless of their origin or identity. An individual's belonging to a particular ethnic or racial group, could, however, be helpful in deciding whether to screen for the specific variant, based on the probability of individuals with a certain geographical background.

Work like Yaron's that involves a molecularization of ethnicity should be understood with particular attention to the racial diversity of the broader context.

ORIGINS OF THE BIOBANK

Diversity is literally valuable for the biobank. The NLGIP is situated in the context of a diverse society, and the NLGIP acknowledges the diversity of the Israeli population. Following Israel's establishment, Jews have emigrated from countries as diverse as Georgia, India, Iraq, Iran, Turkey, and Yemen, in Asia, as well as Algeria, Libya, Morocco, and Tunis, in North Africa, and, more recently, Ethiopia. This diversity has made Israel unusual in its genetic makeup. Other countries, like India, China, Brazil, and, to some extent, the United States, also have an exceptional mix of varied populations from diverse ethnic backgrounds (Gurwitz, Kimchi, and Bonne-Tamir 2003), but Israel is in some ways unique, with many different immigrant groups in a very small country. Also, the Jewish prohibition against intermarriage with non-Jews has led to hundreds of years without much admixture. This makes Israel a "living laboratory" (4).

The NLGIP was established in 1994 as a material resource for studies of human genomic variation. It was established with a grant from the Israel Academy of Sciences and Humanities to serve as the national human cell line and DNA research biobank of Israel (NLGIP 2019a). The NLGIP is located at the Sackler Faculty of Medicine at Tel Aviv University campus, in Ramat Aviv, a wealthy suburb a few miles north of the city center. The lab consists of a biobank of human cell lines and matching DNA samples of donors from the "Israeli populations" (Gurwitz, Kimchi, and Bonne-Tamir 2003, 2). Representing the large ethnic variation and unique nature of those populations, the "NLGIP focuses on collecting, establishing and maintaining human B-lymphoblastoid cell lines and matching DNA samples from healthy donors representing the various Jewish and Arab ethnic groups in Israel" (Gurwitz, Kimchi, and Bonne-Tamir 2003). The activities of the lab are supervised and approved by Tel Aviv University's Institutional Review Board.

The repository consists of human DNA samples and immortalized white blood cell lines, making a collection from over 2,000 donors (NLGIP 2019b), representing the large variation of Israeli populations. The donors include unrelated individuals from diverse genetic backgrounds, including those with European, Asian, African, and Middle Eastern Jewish ancestors, as well as Arab groups: Palestinians, Druze, and Bedouin (Gurwitz, Kimchi, and Bonne-Tamir 2003, 2).

An internationally networked institution, the NLGIP is affiliated with the Pharmacogenomics Knowledge Base, funded by the US National Institutes of Health, and the European Biobanking and Biomolecular Resources Research Infrastructure. It has contributed to the French CEPH Human Genome Diversity Cell Line Panel and to the US-based Coriell Institute's cell repositories (Gurwitz, Kimchi, and Bonne-Tamir 2003). Moreover, the NLGIP is a member of the EuroBioBank,[2] a network of worldwide biobanks that provides human DNA and tissue samples for the scientific research community working on rare diseases. EuroBioBank is the only network specifically dedicated to rare disease research in Europe. It has about 130,000 samples available, which can be requested via an online catalog (NLGIP 2019b). The NLGIP also accepts requests from around the world for samples, which are mailed to researchers at a modest cost.[3] In this way, Israeli citizens' samples become part of the system of global human genetic research.

THE MATERIALITY OF THE BIOBANK

The NLGIP consists of the laboratories (assemblages of spaces, tools, and people) that manage the biobank and the sample storage facilities. The biobank storage area consists of a room containing several round padlocked steel flasks that are kept full of liquid nitrogen. There are also two smaller backup vessels, which are temperature-controlled and alarmed to ensure against accidental thawing and damage. The biobank also contains DNA samples from donors, which are stored in a regular freezer in the cell culture lab, since DNA is stable in water at 4 degrees centigrade.

During a meeting with David, I asked about the biobank, how it works, and what it is. He proceeded to tell me about it at length. Given what I knew of current trends in precision medicine, I asked David if he wanted to sequence the DNA of the biobank's samples to correlate the medical histories of the donors with the genetic profiles of the individuals. This, I thought, would be the obvious next step to gain a better understanding of the genetics of the populations and the relationship between genetics and disease. He said, "No, it would be too costly." I then asked if he would be able to put the medical data online, even anonymized, so that other researchers could analyze it. He said that would not be possible because of confidentiality issues. Each individual sample has an identifier code, which corresponds to a file in a set of folders that are securely stored in David's office. It would not be ethical to publish private data. Occasionally researchers that have been working with samples that the biobank has sent out find a rare mutation and do want to contact the patient to find out about their medical history, but David does not allow them to do this.

I asked David what kind of medical information he has on the participants. "It's not rigorous medical data," he told me. Rather, just a consent form with age, height, and weight (to calculate BMI), smoking habits, any volunteered named chronic conditions, and a very brief family history.

David noted that there was a lot of interest in the biobank in the 1990s and early 2000s, but that they stopped collecting samples two years ago. He referred me to a chapter written specifically about the biobank, "The Israeli DNA and Cell Line Collection: A Human Diversity Repository," which was published in a volume titled *Populations and Genetics: Legal and Socio-Ethical Perspectives* (Knoopers 2003). I relay some of the details here.

In the 1990s, there was a call for a worldwide survey of human genetic diversity (Gurwitz, Kimchi, and Bonne-Tamir 2003) initially by the Human Genome Diversity Project (HGDP). The main arguments were that the initial Human Genome Project—the effort to sequence the first human—would not do justice to the human diversity of the world and that there was also a need to better understand the varying degrees of human susceptibility to disease and historical migrations. The NLGIP was therefore established in 1994 "in light of the awareness to the subject of genetic diversity since the 1950s and the incoming of the Jewish immigrants, and under the influence of the HGDP ideas." The biobank is thus self-imagined as a participant in a global effort to characterize human genetic diversity. In other words, it is a project of global genetic comparison. The laboratory was initially funded by the Israel Council for Higher Education and "was established under the auspices of the Israeli Academy of Sciences and Humanities" (Gurwitz, Kimchi, and Bonne-Tamir 2003, 5). This is the reason, David told me, that it is called the "National Laboratory," the initial sponsors being a national scientific organization. It was not, therefore, an intentionally nationally motivated project in the ethnic sense.

The NLGIP was initially located at Tel Aviv University and was first headed by Professor Batsheva Bonne-Tamir, then the head of the Shalom and Varda Yoran Institute for Human Genome Research at Tel Aviv University. The laboratory was envisioned as a "national repository for human cell lines and DNA samples representing the large variation of Israeli and several Middle Eastern populations" (Gurwitz, Kimchi, and Bonne-Tamir 2003) (see table 3.1).

The NLGIP has a clear ethical policy on its website. It states that "participating researchers must always respect the humanity of the sampled

Table 3.1
Catalog of DNA samples and cells of the NLGIP

Ethnic group	Number of unrelated donors
Jewish	
• Ashkenazi (Central European ancestry)	466
• Ethiopian	72
• Georgian	24
• Iranian	76
• Iraqi	103
• Kuchin (India)	85
• Libyan	89
• Moroccan	150
• Sephardi (Turkey and Bulgaria)	166
• Tunisian	29
• Yemenite	159
Bedouin	58
Druze	79
Palestinian	117

All donors are healthy adults. A matching B-lymphoblastoid cell line is available for each DNA sample. Accessed online, November 22, 2015, at http://www.tau.ac.il/medicine/NLGIP/catalog.htm.

individuals and the cultural integrity of the sampled populations"; that "informed consent must always be obtained from sampled individuals (or their parent/guardian)"; that "the confidentiality of the sampled individuals must always be protected"; that "researchers must strive to avoid misuse of the collected data"; and that "researchers should actively seek ways in which participation in their studies can bring benefits to the sampled individuals and their communities" (NLGIP 2019d). By conjoining respect for humanity with anonymity, the biobank's policy emphasizes that it is to serve the world's people in the most general way and return benefits to both specific individuals and communities. The biobank thus is both a goodwill effort to bring about health improvements for humanity in general, and, in doing so, it also imagines a shared common humanity in the form of a global scientific community of participation. Its goals are beyond any single ethnonation, and it instead provides biomedical knowledge of diverse populations.

COLLECTING OF SAMPLES

Samples donated by patients or healthy individuals were collected in clinics around Israel and prepared for storage at the biobank. Blood cells prepared from the donors consist of B-lymphocytes (white blood cells). These cells may be used for the study of gene expression, for measuring mRNA (the chemical signal that causes the cells to make a certain protein) and specific protein levels, or for specific assays, such as genetic, biochemical, and cell biology studies. Some studies include gene transfection, the alteration of phenotypic properties of the cells, or measuring the effects of hormones or drugs (Gurwitz, Kimchi, and Bonne-Tamir 2003). Samples collected do not immediately fall under the full control of the biobank. Contributors may limit the distribution of their cell lines, demand a request before any transfer of the contributed cell lines is made, and benefit from a free backup service, whereby the biobank holds a secure sample of the tissues donated (NLGIP 2019c). Cell lines donated are intended for *research purposes only*, and anyone interested in using cell lines for commercial endeavors must obtain written consent from the individual contributor first. However, the personal details of the cell line donors remain confidential.

All of the donors are adult Israeli citizens, 18 years and older, "that have given written informed consent for the study of their genetic material (DNA or cells) for biomedical research" (NLGIP 2019b). The listed ethnicity of samples was "self-defined by the donors" (NLGIP 2019b). For the self-defining Jewish donors, their geographical background "is defined according to the place of birth of their four grandparents." The website states "for example, Iraqi Jewish donors have four grandparents who were born in Iraq" (NLGIP 2019b). The gender is also available for all donors, and some data for the age of participants.

The benefits accorded to participants are therefore the understanding that they are contributing to the advancement of science and medicine and that they could benefit from medical experiments on their individual samples later on if they develop a specific disease that could be better treated through experiments on the cells or DNA donated. Donors are not afforded an immediate benefit. The NLGIP acknowledges that the establishment of the biobank raises ethical issues about participation and remuneration, but reports "the IRB prohibited any imbursement (monetary or other) in return for blood sample donations, demanding they must be donated on a full voluntary basis" (Gurwitz, Kimchi, and Bonne-Tamir 2003). In order to increase the number of voluntary, unpaid, donors, the NLGIP thought that individuals already undergoing a routine blood test might be willing to give an additional tube for the biobank. So, blood samples were collected routinely with permission at community clinics in Tel Aviv and across Israel. Druze samples, for example, were obtained from the Carmel region of Israel, while Bedouin Arabs' samples were taken from the Negev region, in the south of the country (9).

USE OF BIOBANK SAMPLES

As mentioned, the biobank at the NLGIP has made samples available to researchers around the world for scientific research. Requests for samples can be made through the NLGIP webpage. All of the order requests received by the NLGIP so far have been for unrelated individuals. I found it surprising that, among those requesting samples, there is such a lack of research interest in ethnic-specific groups. This preference for unrelated individuals, I would learn, is an example of "the more intensive interest of the ordering researchers in human genome variation studies, such as allelic distribution of polymorphic genes across various ethnicities, as well as looking at mutation frequencies and looking for new mutations, rather than more elaborate human genome research, such as haplotype distribution analysis" (Gurwitz, Kimchi, and Bonne-Tamir 2003, 9).

This is to say that the predominant interest in the biobank is not an interest in ethnic genetics, ethnic origins, or ethnic-associated diseases per se. Rather, it is the variation within groups that has interested researchers. The most frequently ordered samples from the biobank DNA, however, are for Ashkenazi Jews (~40% of all DNA samples requested). This is somewhat surprising, since approximately one-third of the NLGIP samples come from families within specific and designated Israeli and Middle Eastern populations, making it a good resource for data about specific Middle Eastern populations, their genetic structure, relatedness, and historical origins. These issues are apparently less interesting to genetics researchers around the world who prefer to work on mapping disease variants within single ethnic groups.

In addition to blood samples, which are made available to researchers for experiments, matching genomic DNA samples can be requested for each of the cell lines (Gurwitz, Kimchi, and Bonne-Tamir 2003). DNA is preferentially ordered over cells, as it is easier and cheaper to ship pure DNA than blood, which can carry viruses and is more easily damaged.

Many scientific studies have emerged as a result of the samples sent out. A 2018 Google Scholar search for "National Laboratory for the Genetics of Israeli Populations" revealed that 260 academic publications have referenced the biobank, with a steady output of articles referencing the biobank each year since its inception in 1994. Perhaps the most noteworthy research to emerge from the biobank samples in relation to the genetics of Jewish identity is a study led by the University of Arizona's Michael Hammer and colleagues, who studied haplotypes (genetic markers) constructed from Jewish Y chromosomes. They traced the paternal origins of 1,371 males from Jewish ethnic groups and non-Jewish groups from similar geographic locations (Gurwitz, Kimchi, and Bonne-Tamir 2003, 12). The study investigated whether Jewish Y-chromosome diversity revealed a common Middle Eastern source population or whether Jewish Y chromosomes reflect mixture with neighboring non-Jewish populations (Hammer et al. 2000). The study concluded that, despite their long-term displacements and movements in different countries, and despite isolation from other Jewish groups, "most Jewish populations were not significantly different from one another at the Y chromosome genetic level" (Gurwitz, Kimchi, and Bonne-Tamir 2003, 12). Such studies are thought to be valuable for their ability to inform historical studies of Jewish migrations and relationships between Jews and host populations. The political import of this work is that it emphasizes the genetic relatedness among diaspora Jewish groups in a way that supports the imagination of the bio-nation. A second major study that the NLGIP provided samples to was the Human Genome Diversity Project (Gurwitz, Kimchi, and Bonne-Tamir 2003). The HGDP was a global consortium with the aim of understanding the varying degrees of human susceptibility to disease and historical migrations.

Other studies describe the Jewish population structure by mapping the matrilineal genetic ancestry of the Jewish diaspora (Behar et al. 2008), by analyzing mitochondrial maternal DNA for evidence of a genetic bottleneck in the early history of the Ashkenazi Jewish population (Behar et al. 2004), and for reconstructing patrilineages and matrilineages of Samaritans and other Israeli populations from Y-chromosome and mitochondrial DNA (Shen et al. 2004). Other work looked at the distinctive genetic signatures of Libyan Jews (Rosenberg et al. 2001). The lead authors on these articles are based at universities in Haifa and Palo Alto.

Although the biobank is housed in the Sackler School of Medicine, research on Jewish genes and the genetic basis of disease does not primarily use the genetic data available through the biobank, nor are the labs associated with the biobank the main producers of research on Jewish population genetics. Further, these kinds of population studies that draw on the biobank are becoming less common as the labs move toward computational analysis of genetic databases. Since it still costs a lot to sequence the full genome of each sample, and since the biobank does not have medical records to go with the samples, researchers now prefer to download data freely available from other genomic databases and analyze the relationship between disease and genetics using computational methods. Consequently, the Israeli biobank is being relatively underutilized compared with, for example, the level of genomic analysis that Noam's lab is doing with genetic data downloaded from databases. The availability of those data renders the biobank somewhat dormant, as databases become more accessible and offer greater amounts of data.

From my research, I learned that the Israeli biobank does not *directly* produce a narrative of Jewish biology or genetic descent, nor indeed of any special aspect of Jewish exceptionalism or uniqueness. Rather, it was conceived as a humanitarian resource as part of global biobanking efforts to categorize human genetic diversity. The biobank was established in the period of relative optimism that followed the 1993 Oslo peace negotiations, and the inclusion of a diversity of Arab and Jewish ethnic groups in the register may have been influenced by a more hopeful moment for the possibility of binational coexistence. Although the biobank resources have been used to bolster historical narratives about Jewish migrations, these studies cannot be traced to the explicit intentions of the biobank's founders. The biobank has been used by other researchers to articulate an imagination of a genetically descended Jewish people, a bio-nation, as for example in the famous "Cohanim" study by Skorecki et al. (1997). However, the Israeli biobank does not engage in outreach or public demonstrations, nor is it open to visitors. Its material resources remain sequestered in an inaccessible and unadvertised room on the seventh floor of the Sackler Medical School building.

The facts that the biobank primarily receives requests for samples taken from nonrelated individuals and that these requests do not usually specify ethnicity suggest that the biobank is not usually used for research on the history or origins of the Jewish people. If the Israeli biobank fosters a sense of moral community, it appears to be a secular humanistic community based on global participation in the advancement of health-care opportunities for humankind. It does not function as a representational space that broadcasts a demographically inaccurate image of the population of the state of Israel.[4] Even while the precise borders of Israel are contested and undefined, the representation of ethnic groups in the NLGIP roughly corresponds to the ethnic makeup of the population under internationally recognized Israeli sovereignty. This is not to say that the biobank is ideologically innocent, or politically neutral, or that it can be understood independently of the territorializing project of Zionism, but rather that it reflects a historical status quo and does not seek to challenge it. If the biobank's value is not so much about establishing historical facts regarding Jews in the land of Israel but rather is about genetic variation across populations in relation to health and disease, this raises the question of the value coalesced in the national biobank. How is the biobank's future value related to the broader landscape of bioscience research and biomedical development?

The significance of discovering variants that are related to the development of specific diseases cannot be fully explained without recognition of the wider political economic context of molecular medicine and especially the move toward personalized therapies. This fact directs attention to two anthropological concerns that were discussed in the lab, namely, the protection of personal human genetic data and the question of the value of genetic data.

REGIMES OF VALUE AND COMMODIFICATION

During my first meeting with Noam, he told me that the Edmond J. Safra Center for Ethics at Tel Aviv University had awarded him a grant to work with a lecturer in computer science, Eran Toch, to study the protection of genetic data. The project was to be conducted by a shared master's student, Netta Rager, who would spend the year writing a master's thesis about methods of protecting privacy in genetic research. Since I was already affiliated with the Safra Center for Ethics as a visiting fellow and had joined Noam's lab as a visiting scientist, he invited me to join the project meetings and offer my perspectives. I told Noam that I would be very interested in participating, and he immediately swiveled in his chair and wrote a short two-sentence email to Eran, suggesting that I join them at their next meeting.

During a project meeting in October 2015 with Eran, Noam, and Netta, I learned about the details of the project and how they were trying to develop a better way of protecting genetic data. The meetings usually began with Netta presenting her work in progress, followed by a group discussion. The three scientists were not, however, strictly interested in protecting privacy outright. Rather, they were primarily interested in coming up with a so-called sensitivity score for specific genes of interest. A sensitivity score would be a way of ranking which genes are most important when it comes to disease risk and rating which genes are not that important. This was considered a necessary and somewhat urgent step in building tools to protect genetic privacy, since the speed and efficiency of sequencing were advancing quickly. This need is not unrelated to global market dynamics.

The economic valuation of genetic data is a consequence of advances in technologies that can sequence DNA faster than before. This process is also expedited by the declining cost of sequencing a person's complete genome, which has plummeted from $100 million in 2001 to $399 in September 2015 (Rager 2015, 3). This figure is expected to continue to drop in the coming years. Moreover, the amount of useful information that can be derived from a person's genome is expanding as the interpretive capacities of data analyses develop. The commonly held notion is that with better databases of genetic information and medical history, scientists will be able to identify the genetic causes of disease and intervene accordingly. But as sequencing speeds increase, and as interpretive power grows, the question of how to protect individuals' genetic privacy arises. Researchers are thus investigating ways of protecting individuals' data privacy, and ethicists and anthropologists are discussing how genetics can impact human identities. These ethical issues were central to Netta's project, and I was able to participate in the project as an anthropologist who could contribute critical perspectives on the idea of privacy and how to think about individual and collective concerns.

The project was to be conducted like this: an online website called the 1000 Genomes Project makes genetic data publicly available. The 1000 Genomes Project website states that the goal of the project is to "find most genetic variants that have frequencies of at least 1% in the populations studied" (1000 Genomes 2015). It is known that the average exome (a person's "coding" genes) still contains more than 13,000 single-nucleotide variants. About 2% of these variants are predicted to affect the encoded protein. By identifying the role of variants, their influence on health and disease may be better understood (Eisenstein 2015). This goal, it is often argued, can be achieved by sequencing many individuals. Since it is still too expensive to fully sequence all of the samples available, data from many samples are combined to allow better detection of variants in any specific region. This way the project can detect most variants with frequencies as low as 1%.

Despite the name 1000 Genomes Project, the project actually has about 2,500 samples (1000 Genomes Project 2015). The website offers VCF files (variant cell format), which contain the list of single mutations or variants in an individual's sample. In contrast with other available genomics databases that offer the medical history of the donors, 1000 Genomes Project does not have information about where donors live, their race, or their medical history. The idea for Netta's project was she would download the VCF files for ten or twenty individuals and identify the rare variants. She would identify the genes that contain the variants and calculate the association of the variants with the reference populations. By identifying the genes that contain the rare variants, Netta could then search for medical information about the importance of the specific genes in diseases and longevity.

By quantifying the "sensitivity" of a gene by the amount of its importance in relation to longevity based on the number of published articles that mention the gene specifically, she would be able to score genes and accord them a specific sensitivity in terms of the medical importance of knowing about it, as well as determine the ethical sensitivity of the gene in terms of public disclosure. Noam thought this project could be extended to make it possible to calculate the cost of treating specific diseases associated with certain genes and ultimately would be able to yield a tool for calculating the likely "cost" of having a specific variant in terms of the expected utility costs of medical treatment. This would be the cost of the risk that the individual would bear as an insured customer, similar to the way that car insurance is calculated by known variables, like gender, age, race, or postcode.

This project is part of a turn toward the commercialization of medical and genetic data. For example, an Israeli company called InnVentis states on its website that "biomedical R&D needs a paradigm shift," since today, "50% of drugs fail in late stage development due to lack of superior efficacy." Meanwhile, the cost to produce a new drug increases (by billions of dollars) and the time to market remains ten years or more on average. They conclude, "the current R&D model is broken" (InnVentis 2015).

To solve this problem, InnVentis envisions that in the future new drugs will be marketed only to the right patients, those who would respond well to the drug. It plans to use mass molecular diagnostics technologies, such as genomics, proteomics, and metabolomics, in combination with big data analytics, which it claims will usher in a new generation of actionable knowledge about disease mechanisms. This new knowledge would improve diagnostics and drug discovery but would also boost the short-term profitability of pharmaceutical companies and the data analytic stakeholders involved. InnVentis expects this new "paradigm" to have "disruptive potential" for the biopharmaceutical and health-care industries (InnVentis 2015).

The context of Netta's project is thus a moment when personalized genetic data are about to become a site for wholesale commodification, as disease state, risk, and treatment populations will be made and reconfigured. In this context of precision medicine, we see individual genetic data engulfed by the valuating context of neoliberal economic rationality, with the condensation of reified value in genetic data and the foregrounding of the individual, the bearer of genetic data, as the rational defender of "private" property.

Let us, however, look at the positive side of what this work is achieving. Specifically, the work I observed in Tel Aviv is helping to improve the treatments for breast cancer and depression.

PERSONALIZED MEDICINE

Noam Shomron gave a talk, "Big Data and Genomics: Halting the Spread of Breast Cancer," as part of Nano World Cancer Day 2017, held at Tel Aviv University in February 2017. In his talk, he presented research his lab had conducted on breast cancer tumors, in which the lab succeeded in stopping tumor metastasis using a novel method. He opens the talk by introducing the field of genomics with the problem of data management. He tells us we have moved away from an era when genomic information was extremely expensive and inaccessible to individual patients or consumers. In 2001, the cost of sequencing a single person's genome was equivalent to the price of four jumbo jet airplanes, whereas today the cost is about the same as that of a bicycle (see figure 3.2).

The avalanche of data that these developments have yielded allows for precision medical therapeutics, which Noam describes as "comparing the DNA of one individual to another individual and trying to fit a treatment, or a therapy, or trying to fit a particular drug, based on these DNA differences" (Shomron 2017b). In the research that Noam presented, the team took genes involved in the development of cancer and, more specifically, the genes related to cytoskeleton organization. These are genes that code for proteins that are essential in cell division, a process crucial for cancer

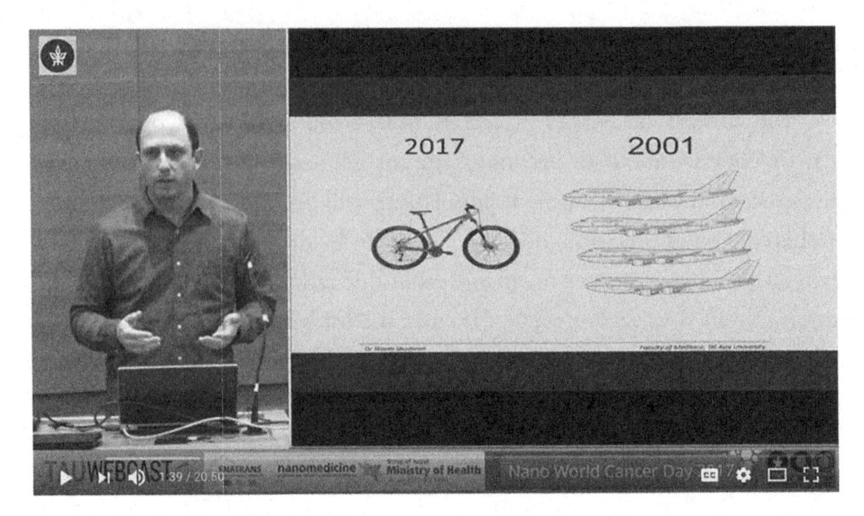

Figure 3.2
Cost of sequencing a single person's genome. In 2001, the cost was equivalent to the price of four jumbo jet airplanes, whereas today, it is roughly the cost of a bicycle (Shomron 2017b).

growth and metastasis. They investigated the 19 genes that could be key targets for stopping tumor metastasis and, in particular, the palladin gene, which, they demonstrated experimentally in vitro and in vivo, can be targeted to block metastasis. They demonstrated this using gold nanoparticles that delivered micro RNA that blocked the palladin gene and therefore stopped the metastasis of cancer cells in the mouse model.

The significance of these results is that, in the future, this technique could be used to stop the spread of breast cancer during the removal of the primary tumor. These results also demonstrate the utility of computational genetic screens, which can help identify genes to target for specific purposes. Although this innovative work is not related to the specific tailoring of cancer treatment for different patients, the presentation of the results was nonetheless framed in the language of precision medicine. It is arguable that the wider banner of "precision medicine" serves to broadly signify future-oriented research that promises therapeutic breakthroughs even if the promise does not fit perfectly with the definition of precision medicine that Noam initially offered. Other research I observed in Tel Aviv did, however, deal with the tailoring of therapies to fit individual genetic profiles.

Another way that I integrated into Noam and David's labs was by participating in an ongoing project that seeks to identify the molecular basis for resistance to antidepressant therapy. The project has to do with one of the main lines of treatment for depression, SSRIs, the most commonly prescribed form of antidepressants. SSRIs effectively reduce symptoms for only about 60% of people with depression. Further, they provoke different sets of side effects in different people, and they take three to four weeks to begin working (Tel Aviv University 2013). Though it is not currently known why some people respond to SSRIs better than others, SSRIs are thought to work by blocking the reabsorption of the neurotransmitter serotonin in the brain, which boosts serotonin signaling and raises mood levels.

David and Noam's labs discovered a gene that they think could reveal how and why patients may or may not respond well to SSRI treatment. Ultimately, this research could lead to a genetic test that would allow doctors to provide personalized treatment for depression and be better able to decide on a course of treatment for each patient. Keren, working with Noam and David, did experiments on cell lines obtained from donors. White blood cells, which are part of the body's immune system against infection, also use serotonin in their signaling, even though they are not brain cells. This makes white blood cells a good model for testing the effect of drugs that modulate serotonin signaling.

Keren had analyzed the RNA profiles of the white blood cell lines that were most and least responsive when treated with SSRI antidepressants. She found a gene called cell adhesion molecule L1-like (CHL1), which was found at significantly lower levels in the most sensitive cell lines and at higher levels in the least responsive cell lines (Tel Aviv University 2013). She also found a new protein involved in SSRI response called integrin subunit beta 3 (ITGB3). She found that the SSRI drug paroxetine caused increased production of the gene for ITGB3. Cell biologists think that ITGB3 interacts in some way with CHL1, but the mechanism is not well understood. By figuring out how CHL1 interacts with ITGB3, Keren would better understand how cells respond to SSRIs and could improve their efficacy by modulating the genes for CHL1 and/or ITGB3.

David had suggested that I talk to Keren and try and figure out a way to measure the interaction between CHL1 and ITGB3. It wasn't clear what experiment could answer this question. I read the relevant research articles on the topic in advance, and I scheduled a time to speak with Keren. We sat on high stools at the lab bench and discussed the project, scribbling potential cell biology mechanisms and theorizing possible experiments that could test the hypotheses. I thought that the use of a fluorescently labeled serotonin transporter could be used to quantify the response of the cells to SSRI treatment. I suggested that we use the gene for a fluorescent-tagged serotonin transporter to measure the effect of ITGB3 and CHL1 on SSRI activity. Keren hadn't used such a technique before but was happy to try the experiment since the tools were available in the lab to do it. I emailed a colleague in Germany (a professor whose class in neurobiology I took at college in Dublin); his lab technician obliged and mailed us a tiny plastic tube with DNA of the fluorescent-labeled serotonin transporter, which we would use in our experiments. It arrived in a padded envelope two weeks later.

This project was ultimately disrupted, as Keren left on maternity leave in April 2016, and I left the lab before her return in August 2016. The core lesson from this vignette, however, is that the research I was involved with centered on genetic markers for disease risk or differential sensitivity to pharmaceutical therapies. The role of racial categories was not an output of the research but rather an a priori assumption, which did not feature further in the discourse or outputs of the research. Racial or ethnic categories would matter in the context of this type of research only insofar as reference populations could be a source of unique variants that associate with a particular condition or pharmacological profile.

This work is typical of research in molecular genetics in Noam and David's labs in that it is unrelated to ancient Jewish or ethnic origins. Such studies point to the importance of biobanks to biomedical research and highlight biobanks' growing importance in the shift toward precision medicine.

BIOBANKS AND NATION-BUILDING

Biobanks have been a growing phenomenon worldwide, especially since the 1990s, when genome sequencing began to provide the possibility of representing the genetic data of large numbers of people. With the advent of fast genomic sequencing, there was much excitement about revealing how many diseases may be associated with single-nucleotide polymorphisms, individual genetic mutations. It was thought that, by identifying the molecular basis for many diseases, a new age of disease prevention and treatment would arrive. It was widely believed that performing genome-wide association studies with the masses of data generated from thousands of individuals would identify many clear-cut disease biomarkers.

Cambon-Thomsen et al. (2003) argue that large biobanks mark a change in the scale of genomics to an industrial-type work organization, with the use of large-scale platforms. Genetics is becoming a wholesale operation, with massive scaling up of the amount of data and the rate at which it can be analyzed. But biobanks usually collect data on populations in specific regions and aim to recruit participants who are representative of the general population (Chadwick and Berg 2001).

Unlike family-based genetic disease registers or centralized medical records, biobanks catalog participants that are representative of a national cohort (Chadwick and Berg 2001). "National" biobanks collect data from the population of a given region or nation (Kaye 2004). National identities and associated imagined communities consequently become refracted through biobanks and the biomedical developments they promise (Busby and Martin 2006). Biobanks, therefore, appear to invite the articulation of visions of the nation, moral community, and natural peoplehood. Busby and Martin claim that "each biobank has markedly different aims, operational arrangements and regulatory regimes" (2006, 238), where varying "ideas of national interest, identity and heritage [are being] constructed and used in the enrolment of support for national biobanks" (241). Moreover, the knowledge that these interpretive projects produce can impact populations' self-understanding and their ethical and moral relationships to the community (246). This can bolster the idea of being part of a lived imagined community.

Despite their capacity to instill an imagination of national coherence, biobanking projects may also be divisive and splinter groups that live within the national territory, thereby naturalizing ethnic or national difference.

Hinterberger writes that "in one of Canada's first large-scale biobanks, French Canadians, who are understood as a genetically close or homogenous population, are contrasted with what are referred to as 'immigrants' and 'Que'becers from various ethnic and racial backgrounds' in public engagement and consultation forums" (2012, 528). She also writes, however, that "some national institutions engage in what might be called genome nationalism" (542). The Mexican Genome Project, for example, made claims that Mexican genomes belong under national state sovereignty, tying national identity and civic participation with the aims of genomic biobanking and medical advancement (Benjamin 2009; Schwartz-Marín and Cruz-Santiago 2016; Schwartz-Marín and Restrepo 2013; Schwartz-Marín and Silva-Zolezzi 2010).

Historian of science Steven Shapin therefore writes that DNA is "an anti-Modernist molecule: a molecular warrant for all the natural differences the conservative thinker could ever want to identify and insist on—differences between unique individuals, between the sexes, races and nations" (2000). At the same time, DNA is "a Post-Modernist molecule, since fragments of our contemporary expert culture insist that the reflexive condition for believing these things about DNA, or indeed disbelieving them, is ultimately ascribable to the workings of DNA itself, while the knowledge of those workings is an authentic item of our culture." DNA thus becomes a site in which we can imagine ourselves while constructing and dismantling imaginations of collectivity that we electively value. DNA biobanks supposedly achieve a bridging of individual biological complexity with a national or supra-individual type, and they represent the imagined natural aspects of populations, and the limits of belonging.

When I went to work in Noam's lab, I hoped to find out what national imaginary, or moral community, is coproduced with the Israeli biobank. I wanted to know whether a national biobank can indeed be "neutral" on issues of national identity. Even if the Israeli biobank does not directly produce a narrative of Jewish identity, origins, or belonging, I wanted to know how it authenticates or reinforces the existing ethnic identities it mobilizes in its categorization of human difference.

SCIENCE AS IDEOLOGY

The discourse of science encodes normative assumptions about what ought to be. In the world of functional genomics, the pursuit of an improved biomedical future both motivates and sustains ongoing research. Anthropologist of genetics Anna Jabloner aptly writes, "The indigenous imagination of genomics . . . entails a persistent, anticipatory orientation toward the future, a variable, but always taxonomical, politics of human biologies, and an ingrained technological meliorism that subordinates the political under the emerging objective truths of a globally circulating, unmarked techno-science" (2015, 28). In thinking about genetic technologies that impinge on the imagination of ethnicity or nationality, it is not just the promises of the future that are at stake, but also the shared history among citizens, their national identities, and the shared world in the political present. As Abu El-Haj argues (2012), genomics offers a science of the predictive medical future while indirectly yielding a mythology of the ethnonational past. The political present and the possibility of continuity hang on the shared imagination of a shared past and a secure future. Scientific discourse acts as a vehicle for purposive ideological projects while purporting to be value-free, neutral, and apolitical. Ethnicity, too, is never value-free or apolitical.

Comaroff and Comaroff read ethnicity as both "ontological" and "orientational," in that it claims a substantial factuality, entailing a volitional commitment to an essential identity as well as participation in a wider ethnos (2009, 45). Similarly, Dominguez in her discussion of ethnic identity in Israel, and specifically Jewish peoplehood, reads peoplehood as a process of "objectification," by which she explains "the possibility that through dialogue and discourse we may assume, or at least come to believe in, the existence of something whose very existence is, in fact, continually 'created' by discursive acts of signification in which we participate" (1989, 21). The discursive creation of the "nation thing" as a scientific object is at issue in ethnic genetics.[5] Dominguez's insights about the objectification of Jewish peoplehood in Israel are helpful for understanding how this phenomenon is playing out today in the domain of genetics.

Ethnic genetics assumes that DNA sequences bear a legible trace of a particular ethnic essence when ethnic groups are structuring categories in genetic analysis. The relation between the individual and the collective is reified, and this process is rendered a necessary and structuring reading frame for ethnic genetics. These conditions, specifically a national imaginary or racial reference population, are the frame of identification. As Comaroff and Comaroff argue in relation to the social ontogeny of ethnic identity: "It is the *marking* of relations—of identities in opposition to one another—that is 'primordial,' not the substance of those identities" (1992, 51; emphasis in original). A relation between individuals is a priori assumed: the nation or ethnos. It is *assumed* in the psychic sense and also in the social, orientational, sense of the word. But the nation is likewise performed as a commonsense assumption. As a historical a priori, it must be recognized that many of the genetic variants used to establish associations between individuals to constitute populations may not matter at all phenotypically. They may not influence the color or texture of a person's hair, the timbre of their voice, their aptitude for music or sport, or their susceptibility to specific cancers. Consequently, a putative, claimed association with others based on non-phenotypic DNA code is a consequence of a situated reading that imputes "unmeaningful" associations: readings that are unmeaningful in the sense that they are far removed from lived and perceptual experience outside of statistical genetic analysis. They themselves cannot be sensibly experienced in normal quotidian life. They escape aesthetic capture.

Moreover, the fetishization—that is, the attribution of animated history, life, and character to noncoding DNA—overlooks the origins of the epistemic practice of genetic reading itself. Instead, genetic reading, as a hermeneutic process, condenses in the genetic substance an a priori signature, an identity that bears an essence by virtue of its relation to both other individuals and the reader, who is situated in a specific historical context. These "ethnic" readings thus succeed in grouping individuals into biological types by their shared unmeaningful differences, their non-phenotypic variants. The origins of these associations are therefore only "objective" (not just an artifact of people's interests or their beliefs) if we disavow the ontological politics at play, the choices that are made in the process of engineering a reading, and of the wider historical conditions that have rendered the reading conditions—the felicity conditions—meaningful, thinkable, and knowable, in the political present.

In addition to reifying assumptions about the biological basis of ethnic groupings, attributing ethnic identities to DNA samples and sequences, even through robust statistical associations, is a double reification: it also misrecognizes the non-phenotypic (non-indexical) genetic sequences as meaningful by virtue of the codes' contiguous relationship with other coding genes that humans carry with them. Signs of nothing become signs of something unique because of their similar clustering in meaningful ethnic groups. The relationship between coding and noncoding becomes a constitutive dialectic of repetition of insignificance, yielding difference, even though such noncoding sequences may not have a phenotypic effect.

This "reading dialectic" is also a reification of the political imaginaries of the present that render meaningful the categories of ethnicity that can be used to sort individuals into historically associated groups and that can be subsumed within an identity category. To arrive at a closed identity that irreversibly chains material things to abstract ideas, as Adorno (1980 [1966]) argues, is to willfully misrecognize concept as substance. It is to conflate ontology with its particular and historically constituted modes of mediation. Rather, it is only ontological claims that should be accorded ontological status and critiqued accordingly.

To continue to critique the process of the molecularization of identity, then, it is the claims and assumptions themselves that must be historicized and displaced. As to where this reading of ethnic genetics fits with the wider school of critical theory, it should be clear that I am emphasizing the critique of ontological claims through an approach that reads around the problem, that describes the context that produces the contours of the problem, and that evaluates the utility of the epistemic claims. In chapters 1 and 2, I looked at citizenship law, political thought, global science, demography, and trends in global biomedicine. This line of thinking is in keeping with a critical social science that can deliver a better understanding of problems facing society, without advocating a particular solution. I thus intend to support a generous anthropology of science that highlights the possibility of "the otherwise" while also attempting to understand the complex overdetermination that has configured the present.

Ethnic genetics can be a purposeful application of a technical tool in the service of building an image of peoplehood, but it can also be put to use in the service of common humanity, although these two projects may overlap in complex ways. The question of the politics of a scientific practice depends on the public discourses that emerge as a consequence of the scientific output and the ways in which populations are imagined and managed as a consequence.

A NATIONAL RESOURCE?

Noam Shomron's lab at the NLGIP contains a small library stocked with popular science books. It includes a biography of Craig Venter (whose private company competed against the US National Institutes of Health's effort to sequence the human genome), a book by the popular philosopher Richard Dawkins, and several chemistry, biology, and genetics textbooks—books both informative and inspirational. I noticed one book with Noam's name on the spine. Curious, I took it down to examine it. The book, a volume Noam had edited, was titled *Deep Sequencing and Data Analysis* (Shomron 2013). In the preface that Noam had written, he opens with a quote from Irish writer Sean O'Faolain, who writes, "There is only one admirable form of the imagination: the imagination that is so intense that it creates a new reality, that it makes things happen" (2013, v).

The power of imagination is not lost on leaders in genetic medicine. Rather, the place of creative imagination at the beginning of Noam's book suggests that imagination indeed comes first, and only then can worlds be changed. This native insight resonates with the analytic idiom of "sociotechnical imaginaries" (Jasanoff and Kim 2015) and indeed demonstrates that theory is not only useful interpretively, from an etic position, but rather, may be precisely the way in which scientists consciously aspire to motivate colleagues and drive their knowledge communities forward.

In the era of speculative "biocapital" (Rajan 2006), it is easy to be cynical about the promises of science and to dwell on the injustices of health-care access. Do the promises live up to the resources invested in the biosciences? Is our ability to define our health being alienated from our hands? Will we end up overprescribed and overtreated as medicine becomes progressively commercialized? These important critiques ought not to be extinguished by the louder voices of technoscientific promise. But I must be proportionate and temper my ambivalence. Despite persistent inequalities in health-care treatment, access, and outcomes across the globe and indeed within the cities of the so-called developed world, the advances over the past century in our understanding of the basic biological mechanisms that underlie wellness and disease have steadily enhanced human health and longevity. Although these benefits have not been delivered equally to all, the technical possibilities have unequivocally advanced. At this moment, the convergence of engineering, computer science, and the biological sciences has created an opportunity to transform the way health-care decisions can be made.

Biobanking and big data analysis have become key elements in bringing about this important material advancement as well as the entailed imaginations about utopian societies and the malleability and controllability of bodies and their futures. But biobanks must not just be understood as technical arrangements or simply as material assemblages. They acquire their meaning and achieve the imagination of value, utility, and meaning in specific local contexts, and as part of global regimes of power and knowledge.

In Israel I determined that the NLGIP is not a nationalistic project. It does not strive to emphasize biological relatedness among Jews or their connection to a territory. It is not an ethnonational biobank; it is a nationally located biobank with extra-national ambitions and activities. The explicit motivations and goals of the biobank were, initially, to be part of a global trend in cataloging the diversity of human populations in the most inclusive way possible. The moral core of the Israeli biobanking project is, therefore, a humanistic one, resting on an imagination of universal human betterment through biomedical development and research. Once samples are sent to other labs for experiments, the NLGIP has no control over the results or of the way in which they are used for ideological purposes. The biobank itself has no goal to establish or displace identities. Prainsack's study of Israeli biobanks has similarly found that, rather than creating novel identities, "biobank projects are more likely to obtain public support and trust if the concepts and terminologies that materialize in biobank practices correspond with established narratives in a particular society" (2007, 86). Similarly, Siegal found that "a striking absence of antagonism between the goals of science and the public good characterizes Israeli discourse" (2015, 767). In line with these viewpoints, I too found that the NLGIP reflects established Israeli concepts and ethnic identities rather than challenging them.

Ultimately my expectations for the biobank were supplanted: although the work I observed in the lab depended on certain racial or ethnic categories, I could not identify a clear moment when the framing national context swayed the research in a particular direction or became an identifiable influencing factor in scientific reasoning. This is a crucial ethnographic finding that has relevance for the methodology of studying science and society. It also problematizes the idea of a local "site" when studying the globalized discourses of science. I found that the discursive social life of genetics and Jewish identity vastly exceeds the science that underpins it. In fact, it raises the question of whether credible biological science underpins the imagination of genomic citizenship at all. The "National Laboratory," I realized, was somewhat like a genetic Holy of Holies: a hollow, empty symbolic space to which is attributed a powerful truth value, coordinating a set of mythical beliefs about the nature of the Jewish nation. Inside the labs, however, there was no Jewish essence to be found. Not only was there no research focus on Jewish origins or the genetics of the Jewish nation, but the work of the biobank and the labs I visited focused predominantly on contemporary trends in biomedicine and an unmarked global rush to precision medicine.

The potential use of the biobank samples for political ends is, however, outside the control of the custodians of the NLGIP, who cannot determine the outcomes of the research that emerges when they share samples with other labs. The means-ends relationship of the technology of the biobank is therefore not guaranteed, and while the biobank offers a means to further the universal humanistic project of biomedical advancement, it cannot rule out the possibility of eugenic or nationalistic science. And if a right-wing Israeli government develops a genetic database of its citizens, the biobank will unlikely be able to intervene to prevent it. That is to say, the science of populations that the biobank has rendered possible could be abused, and its initial intentions could be undermined. However, as the biobank is now becoming relatively underutilized, and as the research moves toward computational analysis, it may become more of a biological archive.

The use value that is congealed in the current collection of tissue samples may become eroded and depleted as interest wanes and the biomedical community sets its sights on decoding the medical implications of genetic variants. As time passes, it may become more important for understanding the genetics of the first generations of Israelis, whose genetic signatures are available for reading in the future.

The most inspiring visions that abound in the contemporary field of medical genetics exist in relation to the growing discourse around personalized medicine and the development of targeted treatments for diseases. Technical progress and the global market logics that drive it forward are far more dominant in the imaginations, aspirations, and values of the labs I observed than anything that can be considered a Zionist science, colored by a sense of biological Jewishness. The Israeli biobank coordinates and facilitates a mythical discourse of genetic peoplehood, to be sure, but a look inside reveals the psychic and performative character of genomic citizenship. Ethnicity remains an idea. The science itself falls far short.

4 BIOBANKING AND "QATARIZATION"

GENETIC MELIORISM

A poster for the December 2015 Functional Genomics Symposium[1] in Doha captures Qatar's labor dynamics and its ambition in the biosciences (figure 4.1). A white male doctor, stethoscope over his neck, reaches up toward a structure of a double helix of DNA, almost grabbing it in his hand. The DNA structure is represented by atomic balls, which is not an unusual representation in chemistry. But in this case, the atoms have a distinctly shiny, pearl-like, appearance—perhaps a reference to Qatar's past industry of pearling. The double helix is composed mostly of gray pearls, but several black pearls are scattered throughout the structure. It appears the clinician is reaching to selectively remove a group of three black pearls that are part of the structure. It is as though the doctor is reaching upward to remove the pernicious and incongruous black balls and restore the monochromatic purity of the DNA double helix.

The poster's symbolism speaks to the transnational dynamics and genetic meliorism that underpin biomedical research in Qatar. The Qatari population has many inheritable diseases, which have been attributed to a history of tribal endogamy (with an estimated consanguinity rate of ~54%) (Sidra 2014, 47). The rapid changes in lifestyle and work that Qataris have experienced in the twentieth century have been accompanied by a rise in obesity and diabetes. Faced with an endless supply of high-calorie food, the prior risks of malnutrition and exhaustion have been replaced with heightened risk of so-called lifestyle diseases. These factors put pressure on the

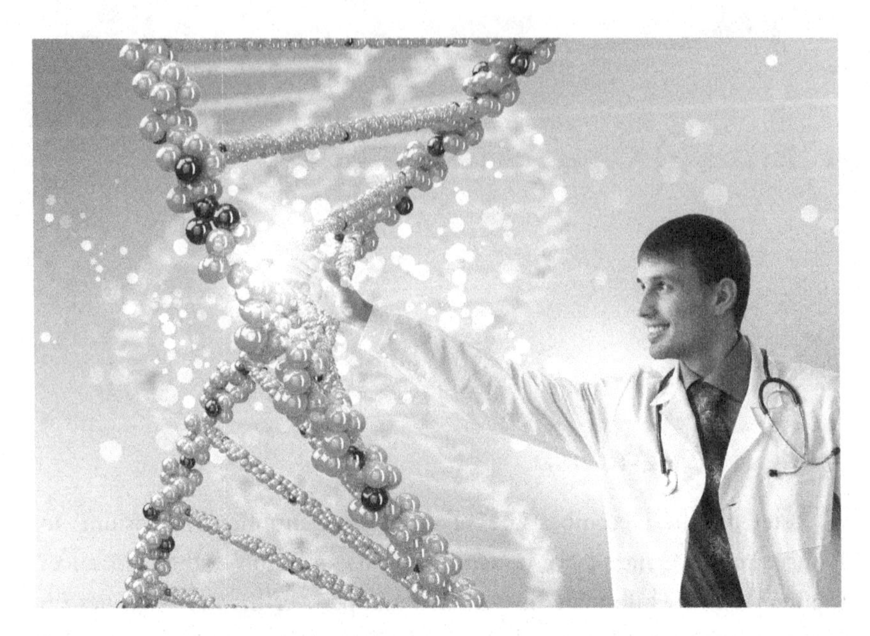

Figure 4.1
Sidra Functional Genomics Symposium, December 2015. The symposium "will offer the opportunity to discuss cutting edge advances in functional genomics and in genomics medicine among world-leading researchers and scientists. The two-day event will feature presentations and discussions that will address the impact of genetic studies on complex disorders and rare diseases. An exhibition from world-leading technology and service providers will run in parallel with the symposium presentations" (Sidra 2015a). Image used under license from Shutterstock.com, © Sergey Nivens/Shutterstock.

state to invest in medical research and health care to make the population healthy and to project an image of a healthy society.

In Qatar, perhaps more than elsewhere, citizen health and national identity are tightly entangled. At the same time that a Qatari national identity is emerging, the Qatari population is being apprehended as a biological object that must also be developed and improved. The new Qatari national identity brings with it the imagination of the "genomic citizen," that is, the citizen is apprehended as a biological object and the target of biomedical improvement. This, we will see, is just one manifestation of a broader move toward nationalization in the Gulf emirate.

FROM PEARLS TO OIL TO BIOMEDICINE

Recent scientific developments have seen many European and American scientists and clinicians traveling to Doha to build collaborations to improve health outcomes for Qatar's population by participating in projects like the Qatar Genome Programme, Qatar Biobank, and Sidra Medicine. These projects are examples of Qatar's purposeful shift from extracting natural resources, namely oil and gas, to a so-called knowledge economy. Qatar has changed tremendously in a very short period of time. A half century ago it was an economically devastated and sparsely populated desert territory; today it is one of the world's richest nations according to GDP per capita (CIA World Factbook 2018). Before World War II, Qatar's main industry consisted of supplying pearls to the world market, but after a crash in pearl prices in the early twentieth century, Qatar suffered from having little else to export and the population endured "years of hunger" (Fromherz 2012).[2] In 1940, the entire population of Qatar stood at just 16,000 (1), when Qatari citizens were enduring extreme temperatures of the desert with little respite. Today they live in air-conditioned and comfortably furnished modern housing. Qatar is the world's largest exporter of liquefied natural gas, and Qataris enjoy a more stable economy than other oil-dependent "rentier states" because of the relative stability of gas prices, in distinction to the volatile crude oil market. Qataris can now spend their time at five-star hotels, malls, and even international universities. Although in the 1950s most oil workers were Qatari (10), most are now foreign, and it is unthinkable for a Qatari to be engaged in manual labor.

As Qatar is a monarchy, and its sovereign leader, the emir, has complete control over the country, the citizens are not formally involved or represented in the processes of governance. This raises the question of how Qataris' sense of nationhood or citizenship is felt, mediated, or performed. What is the national identity or sense of shared community of this relatively young state? First, it must be recognized that genealogy is extremely important in Qatar, where membership in an influential family is a channel to positions of power and influence in the state bureaucracies. Comparing Qatar and its modes of identification to the United States, Fromherz writes,

> One of the first questions that most new acquaintances ask in the USA is what do you do, where do you work? This appears to Americans as the most reliable way of knowing somebody. In many cultures, however, what you are, that is what you are in terms of inherited relations with others, is more important than what you do. Indeed, the extended names of Qataris, "Muhammad bin Khalifa bin Ahmad bin . . . ," for instance, reflect a long string of ancestors rather than the merely one in the case of Western names. (2012, 5)

In Qatar, genealogy matters tremendously in determining where one fits into a specific family-centered, indeed tribal, history of the country. Although in many places the idiom of the "tribe" has become associated with a patronizing colonial anthropology of the "primitive," the "'tribal' lineage is a crucial and internally recognized social form in Qatar." Rather, "ignoring tribes is itself a form of politically correct, neo-orientalism: it means ignoring the major self-identified groupings of Qatar's society, whether imagined or not" (Fromherz 2012, 7). Indeed, each tribe is spatially located in a way that is fairly easy to identify. In each tribal village in the city, regardless of size, "there is a mosque, and a *majlis*. *Majlis* is a term meaning both council and the place a council meets: local meeting-room where qualified men of the tribe decide on internal matters and the relationship between the tribe and others" (21).

Until recently, face-to-face contact between the sheikh and the ruled was not uncommon (Fromherz 2012, 113), and even with the advent of cars, highways, and air-conditioned Western living standards, old alliances and family "bloodlines" have not eroded. Rather, Qatar enjoys what Fromherz calls "neo-traditionalism," a blending of tribal traditions with a modern lifestyle, technology, and urbanization. Even if camels and mud huts have been replaced with luxurious cars and concrete housing, tribal identity remains crucial to one's place in Qatari society. Tribal lineage provides an identity that is arguably more prominent, for most, than national citizenship.

The oil and gas era is not the first time in the region's history that people have depended on natural resources or extractive industry. For hundreds of years, pearling was Qatar's lifeline. Pearling involved almost the entire population, including the nomadic Bedouin who helped guard pearling villages when the divers were away fishing between June and October (Fromherz 2012, 114). There are, of course, differences between pearling and extracting value from natural oil and gas resources. Pearling was a brutally tiring job, and prolonged exposure to the Gulf salt water, coupled with constant heat above 40 degrees centigrade and a long day's diving, could lead to exhaustion and death. Moreover, the divers typically ate only "handfuls of rice and dates" for dinner in order to avoid nausea during the day (117). The rapid growth of Qatar during the past century has seen the risk of hunger replaced with regional political uncertainty.

POLITICAL PRECARITY OR DIPLOMATIC ADVANTAGE?

Qatar's political geography is precarious. Qatar is sandwiched between Saudi Arabia to the west, and Iran to the east, two rival regional superpowers. Since the 2017 "Qatar crisis"—a diplomatic and economic blockade against Qatar—the country has been at the center of a complex web of geopolitical relations involving the United States, Saudi Arabia, Russia, Turkey, the other Gulf states, and their respective positions in the war against the Islamic State. The protagonist in this dispute is Saudi Arabia, who accused Qatar of supporting terror and regards Qatar as being too aligned with Iran, with whom Qatar shares the Gulf's natural gas fields. Qatar's relative economic independence from the other Gulf Cooperation Council states—due to its gas wealth and arguably also its diplomatic ambition in the Middle East region—has led to its political and economic isolation from the other Gulf states. Qatar has long used diplomacy to put itself at the center of many geopolitical disputes. After World War I, for example, the Qataris appealed to American oil contractors to compete against the British, thereby securing greater concessions from Britain (Fromherz 2012, 65).

Beginning in 2005, Qatar began hosting the Doha Debates, a Qatari government-funded platform for international discussion of controversial concerns in world economic and international affairs. Qatar also often functions as a mediator in negotiations between Western states and the Arab states, as well as between Arab states. Qatar mediated between the United States and Libya in 2003, which led to the dismantling of Libya's nuclear program. Qatar allowed the United States to establish a military presence there in advance of the 2003 invasion of Iraq. Qatar has often hosted delegations from the Palestinian factions Hamas and Fatah, and Qatar has facilitated negotiations between Morocco and Algeria (Fromherz 2012, 90).

A foreign visitor might easily get the impression that there are no democratic politics in Qatar. It is true that in Qatar, state power is centered in the person of the emir. Qatar's current emir is Tamim bin Hamad Al Thani (born in 1980), who came to power in June 2013 after his father's (Sheikh Hamad bin Khalifa) abdication. Tamim was educated in Britain and attended the Royal Military Academy at Sandhurst. Tamim's father, Sheikh Hamad bin Khalifa, of the Al Thani dynasty, took power from his father, Khalifa bin Hamad, in a bloodless palace coup in 1995. It is reported that the deposed sheikh had "lost most of the energy of his youth and, according to some reports, had descended into alcoholism" (Fromherz 2012, 85). After seizing power, Sheikh Hamad made it clear that he had a vision for the future and the development of Qatar. He immediately sponsored and hosted the news network Al-Jazeera, started a slow process of democratization and reform of the electoral system (he created Doha's municipal council in 1999) (83), and his wife (the second of three) Sheikha Mozah has since led in cultural reform, establishing Education City (home to several US university campuses) and large-scale biomedical development projects. Despite these steps toward modernization, democratization, and globalization, it should be noted that in Qatar, no meaningful distinction is made between the emir's person and the political and legal organization of the state.

After his father handed over power to him in 2013, the emir became the sovereign power incarnate, giving him the freedom to pursue his family's interests and commercial desires with the resources of the state at his discretion. Moreover, the emir can grant citizenship to whomever he chooses, and he can also take it away without any oversight or process. There are, however, in theory, some limits on the emir's power. In 2003 the state established an advisory council (after 96% approval in a popular referendum), over which the emir has the power of veto that can be overruled by two-thirds of the council. However, the emir appoints one-third of the council members. In cases of the emir being overruled, he can also suspend legislation and delay decisions he disapproves of (Fromherz 2012, 126).

TRIBE OR NATION?

The Al Thani dynasty was established later than that of the United Arab Emirates, Bahrain, or Kuwait, but it had established control of Qatar by the end of the nineteenth century, before the discovery of oil. The family also controlled the area around Doha at that time (Fromherz 2012, 17). Following World War I and the fall of the Ottoman Empire, Qatar was designated a British protectorate. In 1968 Qatar attempted to form a federation of Arab emirates in the Gulf, but after Qatar's proposal of Doha as the capital was rejected, Qatar reversed its interest in being part of such a federation.

In 1971 Qatar became an independent state. By facilitating the Al Thani family's central position of power in the state, Britain also helped to establish in Qatar a monarchical-type dynasty, of which there was probably no tradition in Qatar before an 1868 treaty between Muhammed bin Thani and the British Colonel Pelly (Fromherz 2012, 53). Before the British deal, political authority was defined by family property or religious grounds and determined via Islamic principles and local traditions (57).

Despite the tribal-family nature of the state, a strong national identity is now emerging. The establishment of a national identity involves manufacturing an imagination of a shared historical experience (Anderson 1983). In Qatar, one particular historical event has been grasped as a moment of collective identity formation, a battle against a foreign force that brought Qataris together. In 1892, 200 Ottoman soldiers arrived in Qatar to stake a claim to the territory, and after a refusal by Sheikh Jassim to meet with the Ottomans, these soldiers captured 13 Qatari chiefs. Sheikh Jassim responded and, in unifying and leading a group of Qataris to battle, defeated the Turks at the site of Wajbah. This battle is now annually commemorated as National Day, first celebrated officially in 2007. Since then, National Day has been commemorated annually with parades, military displays, and cultural events. Preparations for the 2016 National Day included an Arabic-language festival.

However, the emir ultimately canceled the 2016 National Day celebrations—just days before the festivities were set to begin—in solidarity with the people of Aleppo, the site of massive destruction and thousands of civilian casualties during the Syrian civil war. National Day, Fromherz writes, "has somewhat surpassed Independence Day, in the size and importance of the celebrations, despite beginning only in 2007" (2012, 61). It appears that the emerging National Day is displacing Independence Day, with the effect of emphasizing national collectivity over tribal particularity. To understand the character of Qatar's emerging national identity, we need to consider how a sense of shared peoplehood is being inculcated.

A NATION IMAGINED

In Qatar the relation between state and citizen is not like it is in Western liberal democracies. In Qatar, the citizens do not fund the state with taxes. Rather, the state supports the citizens financially, in return for their consent to be governed by the emir. Although Qatar's total population is 2.4 million (July 2020 estimate), its demography is heavily composed of non-Qataris (CIA World Factbook 2020b). Males greatly outnumber females at an overall ratio of 3.39 male(s) to 1 female, though this ratio varies by age (figure 4.2) (CIA World Factbook 2020b).

The second half of the twentieth century saw workers flocking to Qatar to help build the country's roads, buildings, highways, and support infrastructure, and today, Qataris are officially estimated to compose 11.6% of the resident population (Business in Qatar and Beyond 2013; CIA World Factbook 2020b). Unsubstantiated rumors that were relayed to me by scientists during my visit in December 2016 suggest that the Qatari population is actually now a significantly smaller percentage of the resident population. This minority status necessarily raises the question of how a sense of national identity can be produced, maintained, or publicly performed in a country in which the citizens are vastly outnumbered. Fromherz writes that although the Al Thani family uses historical myths and heritage to maintain their rule, "tribal affiliation and solidarity is slowly being replaced by national solidarity" (2012, 29). Qatar is therefore in the process of transforming from a tribal, segmentary state toward a unitary state with power centralized. The Al Thani family has not used force to maintain their position but has used the idea of pre-oil Qatari independence to inculcate a sense of solidarity and loyalty to the state (157). By glossing over history, they attempt to turn "tribal affiliation into a sanitized form of 'heritage'" and maintain power over the state (160).

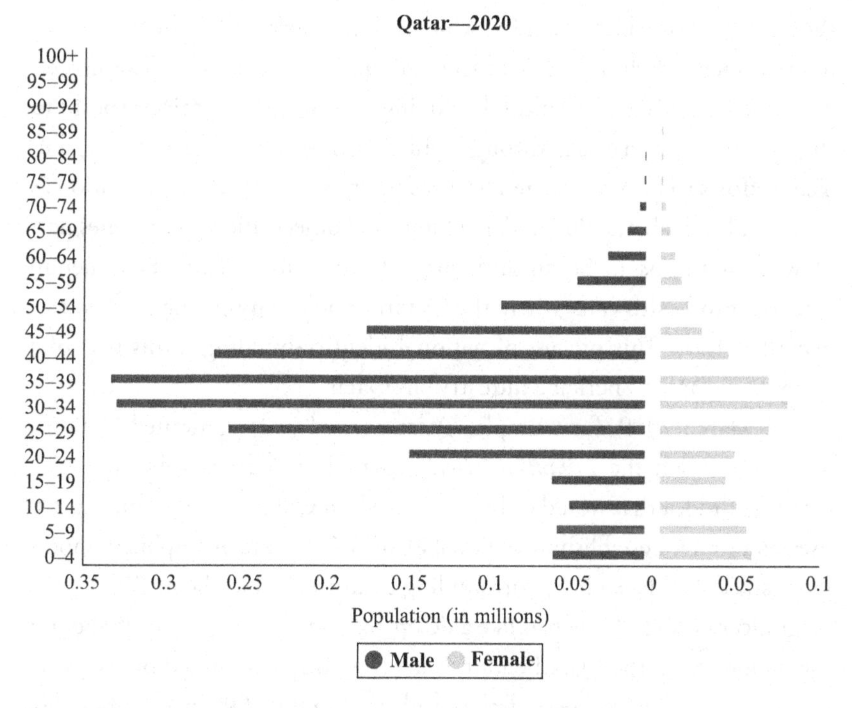

Figure 4.2

Qatar population pyramid. A population pyramid illustrates the age and sex structure of a country's population and may provide insights about political and social stability, as well as economic development. The population is distributed along the horizontal axis, with males shown on the left and females on the right. The male and female populations are broken down into 5-year age groups represented as horizontal bars along the vertical axis, with the youngest age groups at the bottom and the oldest at the top. The shape of the population pyramid gradually evolves over time based on fertility, mortality, and international migration trends (CIA World Factbook 2020b).

In 2016 and 2018 I attended the National Day festivities in Doha at the Darb Al Saai plaza. A host of activities hinged around national identity, health improvement, and Qatari heritage. They included a kiosk promoting organ donation, lectures for children about a healthy diet, blood donation stations, military displays, camel riding, informative posters about agricultural science such as date cultivation, shooting ranges and horse riding for children, and an audiovisual cartoon presentation for children aimed at improving Qatari's Arabic-language skills. Qatar Biomedical

Research Institute had a stand at which children were taught how to examine a sample of their own cheek tissue cells under a microscope. This diverse set of medical, scientific, cultural, and linguistic outreach projects speaks to the multifarious channels through which citizenship is fostered in Qatar. These efforts to improve citizen health simultaneously present the nation as both a cultural abstraction and as a biological object. Biology becomes part of what it means to be an authentic Qatari citizen. This phenomenon extends into health care, where the Qatari population is being studied as a genetic cohort. This process of national identity-building forms part of a larger burgeoning "heritage industry" in Qatar.

In March 2019, for example, Qatar opened its long-awaited National Museum, which the *Financial Times* reported as "an absurdly expensive, extravagant idea that would only ever be built in Qatar, where architecture is used as a tool for establishing national identity (a feature of amplified importance since the blockade by surrounding countries)" (Heathcote 2019). The magnificent building, shaped like a desert rose, serves as a potent symbol for the nation itself. The Qatari nation is inadvertently naturalized by its synecdochic relationship to the landscape. Other examples of the nation rooted in symbols from nature include a project at Weill Cornell Medicine–Qatar (WCMC-Q) that published the first genetic maps of the date palm (L. S. Mathew et al. 2014) as well as the Arabian oryx (Weill Cornell Medicine–Qatar 2009). The oryx is a symbol of the Qatari nation, recognized around the world as the logo for Qatar Airways. And the date is a symbol of Arabian hospitality and a traditional Islamic way of life, as it is believed that the prophet Muhammad broke his fast eating dates. In keeping with the Islamic tradition, WCMC-Q published a related study examining the metabolomics of Ramadan fasting (S. Mathew et al. 2014).

Another major project in the heritage industry is the Msheireb Properties development, a subsidiary of the Qatar Foundation for Education, Science, and Community Development. The Qatar Foundation, a private, independent, nonprofit organization, was founded in 1995 "to support the development of national centers of excellence and position Qatar as a global leader in innovative education and research" (Sidra 2015b). Msheireb Properties is an urban development project in downtown Doha, in the Msheireb area, which is the oldest part of Doha. The new development also features a museum complex, consisting of four "historic heritage houses." The museum brochure states that the houses "reveal unique aspects of Qatar's cultural and social development and inspire to create trusted environments in which the people of Qatar will engage, converse and exchange thoughts about their past and their future." The four heritage houses are Bin Jelmood House, Company House, Mohammed Bin Jassim House and Radwani House. Sheikh Mohammed Bin Jassim Al Thani, the son of the founder of modern Qatar, originally built Mohammed Bin Jassim House. The house has original rooms furnished as Bin Jassim inhabited them in the early twentieth century, so as to preserve a memory and understanding of the heritage of Qatar.

Radwani House is quite similar in that it presents traditional Qatari family life and documents the transitions that have led to contemporary Qatar. It features artifacts from domestic family life, describes how the coming of electricity has affected social life, and displays images and accounts of the Msheireb region in the early twentieth century.

Company House was once the headquarters of Qatar's first oil company, and the museum tells the story of Qatar's petroleum industry through the lives of the workers and families that labored to provide the foundations for the modern nation. The displays include statues of laborers holding tools and engaging in hard physical labor, as well as an audiovisual theatre that plays a short documentary on workers' lives during the early oil years.

Bin Jelmood House is essentially a slavery museum. Its ambitious aim, the brochure states, is to "raise awareness and play a pivotal role in the global abolition of human exploitation." The museum comprises a series of chambers, each with its own audiovisual presentation. The first room presents slavery as a practice that in the past was widespread across the globe, and it relativizes European serfdom with other forms of enslavement around the world. The exhibits follow a chronological order, progressing from the East African villages where slaves usually originated, through the passage to Zanzibar, and through Muscat, to Doha, where slaves' daily lives and integration into Qatari life are presented. The series ends with a number of displays recognizing contemporary cases of human exploitation, such as those involving child laborers, sex trafficking, and the abuse of the Kafalah contract (a system of sponsoring migrant labor based on Islamic law) common in the Gulf region. The museum's brochure puts a positive spin on Qatar's history of slavery, stating that "the story in Qatar begins in enslavement but ends in shared freedom and shared prosperity." Crucially, however, the museum has recently installed a DNA museum titled "Journey to the Heart of Life" as a section in the series of exhibits. The DNA museum demonstrates through several interactive touchscreen activities the uses of genetics in revealing population origins and explains the ways ethnic identity can be explored through genetics. This is just one example of the intersection of genetics and ethnic identity in Qatar.

Another way in which heritage is being deployed as part of the nation-building project is through a project between UCL Qatar and the Qatar National Library, which aims to create a digitized archive of Qatar's early aerial images. There are more than 15 million photographs available, mostly originating from Royal Air Force reconnaissance missions in the region (Qatar Foundation 2016). The project is described by the Qatar Foundation in their monthly magazine as an important national heritage resource and presented as a project that will not only provide a visual archive on Qatar as a territory but also will "map the roots" of the nation. The same issue also features an article about promoting equality among children with diabetes—another example of how medicine and health care are sites for promoting national solidarity and the imagination of an organic and naturally rooted nation.

While the "mapping the roots" project aims to celebrate the territory of Qatar as a sovereign object, this emergent Qatari national imaginary gains an evidentiary footing in specific biomedical developments and capacity-building in the biomedical sciences. In the biomedical sciences, it is the nation itself that is understood as a natural entity and reified as an object of interest.

When I first arrived in Doha in 2015, I had already begun my stint at the National Laboratory for the Genetics of Israeli Populations in Tel Aviv. I wanted to address some of the same questions: What kind of moral community do Qatar's biomedical developments and biobank constitute? In what ways are ideas of natural peoplehood articulated or refracted through these institutions? What kinds of ethnic distinctions are made, or unmade, through the biomedical research on the Qatari population?

SIDRA MEDICINE

In recent years, Qatar has established a constellation of biomedical projects around Doha, and one of the essential nodes is Sidra Medicine. Sidra Medicine, a state-of-the-art academic medical center, is affiliated with the Qatar Foundation. Work at Sidra focuses on three key areas: world-class health care for women and children, medical education, and biomedical research (Sidra 2015b). Sidra represents the vision of Her Highness Sheikha Moza bint Nasser, Sidra's chairperson, and has been designed to become a center of the highest international standard, with the latest medical equipment and laboratories to further knowledge and clinical advancement. Her Highness intends for Sidra to be ranked as one of the most advanced research hospitals in the world, setting new standards specifically in women and children's health care,[3] while also helping to build both Qatar and the Gulf region's scientific expertise and resources.

Sidra has an international academic partner in WCMC-Q, and Sidra will also become a primary teaching facility for WCMC-Q, offering both Qatari and international students the chance to develop clinical skills and participate in biomedical research. Researchers who work at Sidra may have academic appointments at WCMC-Q. For example. Sidra will specifically support investigations pertaining to women's and children's health in accordance with the Qatar National Research Strategy. Through these collaborations, the hope is that Sidra will further the understanding of the genetic bases for common and rare disorders, particularly in the Gulf region, as well as the environmental or biological factors that influence their etiology.

The Sidra complex was designed by American architectural firm Pelli Clarke Pelli. The towering structure of steel, glass, and white ceramic tile was thought to provide the ideal environment for tranquility, privacy, and healing. The building design also incorporates three towering atria that act as "indoor healing gardens"—which each patient can see from their luxurious private room—incorporating soothing water features and adorned with an "impressive art collection" (Sidra 2015b). The Sidra development has not been untarnished by controversy, however.

A set of sculptures by British artist Damien Hirst that depict the stages of human life development, from fetus to birth, including 14 fetuses, were commissioned for the Sidra entrance area. The series chronicles the development stages from conception to birth and is completed with a 46-foot bronze statue of a baby boy. The statues were reported to have cost $20 million (Jones 2013); the *New York Times* reported that "although the sheikha declined to confirm or deny the reported cost of Mr. Hirst's sculptures, she said the outlay was 'not a crazy number'" (Vogel 2013). The concerns that the statues raised were not about the cost, but rather about the exposure of the naked body.

In December 2016, I arrived at Sidra to interview a senior manager, and I was excited to visit the exhibit and take some photographs. When I couldn't find the statues, I asked the reception staff how I could view them. They told me that the statues were covered up and were no longer on display to the public. Surprised, I asked why this was the case. Three staff members at Sidra told me that "the older generation didn't like it." They consider pregnancy and the body to be a private domain, the staff said, and the statues offended their sensibilities. After remaining concealed for some time, the installation, named *The Miraculous Journey*, has been open and on display again since late 2018. Its temporary concealment reveals the tensions at play within Qatari society: the country is fast globalizing, importing elite global art, and challenging older traditions, industries, and moral commitments about modesty and the concealed body. From a distance the scientific development plans of Qatar appear to be proceeding quickly and successfully, but challenges and difficulties arise as Qatar changes socially and culturally. The controversy surrounding the sculptures is an example of how different segments of Qatari society differentially imagine what kind of future they want for themselves.

SIDRA'S IMAGINED FUTURE

The Sidra Medicine development is located at the heart of Education City, which is the main development site of the Qatar Foundation. The Sidra building was provided with an initial research division budget for five years of $709,714,000; Sidra has more than 10,000 square meters of research labs.

Sidra's vision was outlined in its five-year strategic plan (2015–2020), which was published in August 2014 (Sidra 2014), in order to give a sense of the scale and ambition of biomedical development in Qatar. Sidra was launched as part of Qatar's broader development plan, specifically the "2030 national vision to turn Qatar into a knowledge-based society at par with most technologically advanced countries" (3). Although Qatar had already reached a relatively high standard in computing, engineering, transportation, and information technology (IT), medicine and biotechnology lagged behind. Sidra was intended to address the gap with innovative biomedical technologies and outstanding medical care. The project was likewise thought to play an autonomous role in developing and testing novel ideas, in line with the Qatar National Research Strategy and National Health Strategy. Sidra was imagined as integrating into the existing health-care system by cooperating with other Qatari institutions, for example, by providing tissue samples and research materials. It was conceived as open to collaboration within Qatar and as well as with overseas research partners.

Sidra's three key goals are to "prioritize translational research; state of the art training for Qataris; and support clinicians to practice personal medicine" (Sidra 2014). Translational research in this context means the application of basic biology and biomedical research to clinical practice. Indeed, most of the efforts of the research branch were planned to be in the area of personalized medicine. Sidra is building essential capabilities for Qatar in the latest biomedical science, but it is also a part of a conscious nation-building project. The plan explicitly states, "*Qatarization* remains Sidra's ultimate goal in capacity building" (Sidra 2014, 6; emphasis added). Consequently, the development of Sidra will yield a vibrant international academic community and will draw "top scholars independent of their nationality of origin to create and sustain a global community of leaders in the biomedical field" (Sidra 2014).

Qatarization is a government initiative to get Qatari citizens working in the public and private sectors at a higher rate. For example, the *Qatar Tribune* (December 18, 2016) reported that the customs authority has achieved 95% Qatarization. Qatarization does not merely mean a numerical achievement of having Qataris employed in roles but also involves training Qataris in high-level skills, particularly IT, so they can contribute to the technological advancement of the state. Qatarization is thus a policy of both skilling and promoting Qataris.

The flow of foreign migrant workers to Qatar has posed a challenge to state initiatives like Qatarization. In some industries, like the biosciences, Qatar is drawing top talent from around the world, potentially reducing the industry's Qatari proportion. It remains part of Qatar's vision to attract expertise from around the globe and bring talent to Qatar. The policies of Qatarization are thus intended to drive Qataris into high-skill roles in the workforce at a higher rate than before, even if this goal stands in tension with current migrant labor practices. In the long run, Qatarization would presumably render Qataris more visible in public life and would help foster a sense of collective nation-building. The national biobank, for example, is a key site where the emergent Qatari national identity is affirmed.

QATAR BIOBANK

Qatar Biobank (QB) is an entity separate from Sidra but it is related in its goals, activities, and funding structure (Qatar Biobank 2015a). QB is a center within the Qatar Foundation, and was created by the Qatar Foundation in collaboration with Hamad Medical Corporation[4] and the Supreme Council of Health, with the broad goal of furthering medical research on Qatari health issues. Its pilot phase began in 2013, and since 2014 it has been in the biobank initiation phase.

QB consists of a collection of samples and information pertaining to the health and lifestyle of members of the population of Qatar and offers research opportunities for Qataris as well as scientists and clinicians from the region and the world. QB aims to become both a resource for Qatar and a globally recognized and competitive institution. It aims both to further research that will benefit the Qatari population and to offer opportunities for Qataris to participate as donors or as professional researchers and clinicians. Its website describes the bank as "a scientific and altruistic partnership between the research community and the people of Qatar to build a better, healthier future for generations to come." Its principal mission is to "act as the Qatar National Centre for biological samples and health information to enable research towards the discovery and development of new healthcare interventions" (Qatar Biobank n.d.-a).

QB focuses on genomic medicine and systems biology, diabetes, cancer, medical genetics, obesity/metabolic syndrome, respiratory health, nutrition, mental health, cardiovascular disease, cognition, and the relationship between health and socioeconomic status (Qatar Biobank 2015b). QB had an exhibit at the Sidra Functional Genomics Symposium in 2015 and 2016, and at the 2016 symposium I took one of the brochures they made available to visitors, which was titled "Planning for the Health of Our Future."

We can presume the deictic "Our" denotes the Qatari self-referential "We" and addresses Qatari citizens as part of both a nation and a biological collective within a national temporal frame of progress toward improved health and indeed a healthy, prosperous future. The "sociotechnical imaginary" (Jasanoff and Kim 2015) bridges incontestable biological facts like "the body is made from different tissues, . . . your genome is inside the nucleus," and so on, to knowing "which genetic differences or environmental factors are important" to deliver "new drugs . . . new diagnostic" tests. This could be read as the universal narrative of global science and its promise of progress, but in this instance, it is mobilized within the context of Qatari national development, addressing, indeed interpellating, Qatari citizens selectively.

To this end, QB will be both a platform and a driver of health research as it recruits large numbers of the Qatari population to donate biological samples as well as information about their health and lifestyle. Probably the greatest health concerns for the Qatari population are obesity and, consequently, diabetes. An initial report of QB reported, "17% of our adult population suffers from type 2 diabetes" (Qatar Biobank n.d.-a). QB published findings from a pilot study that addressed the physical activity of Qataris and the dominant reasons for clinical referral (Qatar Biobank 2015c). The study report is based on research between September 2013 and October 2014, with 1,200 samples collected during QB's pilot phase (Qatar Biobank n.d.-b).

The study published the following findings concerning physical activity: 80% of the sample population reported no level of moderate physical activity per week (77% of males and 86% of females reported no moderate physical activity per week); 67% of the sample population (61% male, 70% female) reported walking less than two hours per week for leisure; 55% of the males sampled reported working in an office-based environment, which leads to inactivity; and 42% of the males sampled reported watching TV and using computers for more than four hours per day compared to 38% of females sampled. In relation to clinical referral, QB reports that 373 (70%) of the participants were unaware that they had a disease; 25% of referrals were due to abnormal bone density and low blood calcium rates; 19% of referrals were due to dyslipidemia (high cholesterol); 18% of referrals were due to diabetes; and 17% of referrals were due to high blood pressure (Qatar Biobank 2015c). These data suggest that the Qatari population is not getting enough exercise and that their lifestyle habits may be harming citizens' health.

Indeed, the initial study also reported on the levels of obesity and risk of cardiovascular disease in the Qatari population: 73% of the sample population was classified as overweight; 37% of the population was classified as obese; 37% of the population has borderline or high levels of total cholesterol; and 76.6% of male and 70.4% female participants were at risk of developing cardiovascular disease because of being overweight or obese. Of these participants, 864 are male and 1,142 are female, and all age groups were represented in the sample. However, the majority of the participants were between 22 and 38 years old (Qatar Biobank 2015c). Taken together, these findings make clear a need for improving citizen health and lifestyle habits.

It should be noted, however, that the distribution of samples collected is not proportionally representative of the demographics of Qatar. If it were, both women and Qataris would be in the minority. Rather, it seems, for whatever reasons, the sample is roughly equal between male and female donors, while Qatari participants stand at 2,360, in contrast to non-Qatari participants, who number fewer than 700 (table 4.1). In the demographic sense, the biobank is not a representative assemblage of the residents of the territory under Qatari sovereignty.

QB aims to recruit more than 60,000 participants by 2019 (Qatar Biobank n.d.-b). As of September 2018, 15,000 participants have been recruited (Qatar Biobank 2018). Any adult (i.e., over 18 years) who is either a Qatari national or a long-term resident (having lived in Qatar at least 15 years) can contribute to QB. According to figures released in January 2015, QB had recorded 2,006 participants, 1,500 of whom are Qatari, and 506 of whom

Table 4.1
Qatar Biobank participant data

	Number of participants	Number of samples
Total	3,022	234,157
Male	1,594	126,322
Female	1,428	107,835
Total Qatari participants	2,360	158,739
Male Qatari	1,179	86,158
Female Qatari	1,181	72,581

Between January 2013 and January 2015, "3022 participants have provided a total of 234,157 samples. From these figures 2360 of the participants who provided a total of 158,739 samples are Qatari" (Qatar Biobank 2015d).

are long-term residents (Qatar Biobank n.d.-b). To register online for an appointment with the biobank requires a Qatari identification card number. The process takes around three hours and involves the donation of samples of urine, saliva, and blood and undergoing a series of measurements (including height, weight, grip strength, blood pressure, body composition, heart and lung function). Participants also complete a questionnaire. The biobank celebrated its four-year anniversary in February 2018.

QB is imagined as becoming a valuable national resource for Qatari health. As genetic and health information and samples contributed grow in number, researchers will be able to study how lifestyle, environment, and genes affect health and illness. The knowledge produced could help in the development of better medical treatments and disease prevention measures for present-day Qataris or future generations. There is also an important ethnic dimension to the research. The QB website claims that until now most medical treatments have been developed through the study of Western populations and that there has been a lack of large-scale research on populations in Qatar or in the wider region. QB is now one of the region's most ambitious biobanks, and it aims to play a key role in helping prevent and better treat diseases that affect Qatari populations. Further, the knowledge produced by QB, the organizers claim, will lead to tailored health care and personalized medical treatments.

Indeed, QB's pilot report states that it "will chart a road map for future treatment through personalized medicine" (Qatar Biobank n.d.-c, 7). Dr. Asma Al Thani, vice chairperson of QB's board of trustees, states that "Qatar's scientist and research community recognizes the current shift from traditional genomics, as the mapping of an individual's DNA, to the population-based studies institutions across the world" (7). QB thus situates itself at the forefront of genetic research and recognizes the shift from traditional genetics and biobanking to genomic analysis, entailing the search for disease biomarkers leading to personalized treatments. But what kind of population does the biobank imagine when it collects samples and presents public health data? The gender, ethnic, and age distribution pyramid for the initial participants is starkly different from the population pyramid for the whole population (figure 4.2).

The population that is selected and imagined through the initial biobank study resembles a "normal" demographic distribution by foregrounding the role of adult Qatari men and women with a relative balance in the demographic makeup of the biobank. The biobank population pyramid has a wide base that comprises youth, with a decrease in numbers as age rises and a relatively balanced gender distribution. Since Qatar's population is primarily composed of migrant workers, or noncitizens, its demographic structure includes a disproportionate number of men in their twenties and thirties. Whether because of the way the biobank has selected participants or other reasons, only certain elements of the population have had a chance to be represented. Therefore, the biobank may be criticized for failing to provide a sample that roughly corresponds to the territory's population diversity. But reading this phenomenon as a cultural artifact of nation-building, it becomes clear that the biobank is acting as a site through which a "normalization" or, more precisely, Qatarization of the population structure can be imagined. The demographic minority status of Qataris arguably threatens the national character or identity of the young nation. Perhaps this minority status may be assuaged by assembling public health data in a way that presents a picture of the population that foregrounds Qataris and their dominance in the biobank repository. In the biobank the Qatari citizen is apprehended as a biological entity for health improvement, but this process can also help inculcate a sense of national citizenship and community.

MORAL COMMUNITY AND QATARIZATION

The biobank offers a chance for Qataris to participate in the process of producing data as "citizen scientists." Dr. Hadi Abderrahim, managing director of QB, sees the biobank as an opportunity to bring the public in as participants and as partners. He claims:

> Qatar Biobank does not only aim at recruiting the public to take part in biomedical research, but also wishes to partner with our public and help them become "citizen scientists" who, through their personal contributions, play an active role in the process. As such, Qatar Biobank's recruitment approach provides a model for public involvement in biomedical research and promotes Qatar's dedication to raising awareness and commitment, engaging the community in shaping a better health of their future generations. (Qatar Biobank n.d.-c, 8)

The biobank is presented as part of a broader move toward crafting a vision of the good citizen as technologically informed, healthy, and engaged with the global community. Dr. Hanan Al Kuwari, chairperson of the biobank, describes the biobank as "a scientific and altruistic partnership between the research community and the people of Qatar to build a better, healthier future for generations to come" (Qatar Biobank n.d.-a). In his view, the biobank constitutes a relationship between the research community—which presumably extends far beyond Qatar—and the people of Qatar. The biobank is not conceived of as a state entity or as a collective representing Qatari interests. It is unclear, however, whether the above use of "people of Qatar" refers to Qatari citizens, ethnic Qataris, or residents of Qatar. In any case, the goal is to place Qatar as an equal with other biobanking nations and their associated research endeavors. It is also the international biomedical research community that is said to benefit from collaborating with the biobank and not just the Qatari people themselves. Accordingly, the biobank states, in relation to its research, "Qatar Biobank will make it possible for scientists to conduct research to address some of the greatest health challenges facing Qatar and the region, including cardiovascular disease, obesity, diabetes, and cancer" (Qatar Biobank 2015e).

The biobank and the research it will engender are also presented as a great opportunity to the world's scientists. In addressing the specific benefits to the researchers, the biobank states that "the unique breadth and depth of the information and samples collected by Qatar Biobank on the population of Qatar will allow researchers to advance the understanding of local and regional health and disease to enable new and exciting developments in healthcare and medicine" (Qatar Biobank 2015e).

QB is envisioned as a major contribution to the global study of human genetic diversity, even though the project is focused on the local medical needs of the Qatari population, which is in fact a demographic minority. As Qatar moved from pearls to oil, its population experienced a rise in diabetes and obesity as a consequence of a change in lifestyle. Today, however, they are pursued by state policies as the therapeutic subjects of the most advanced health care available on the planet. But a particular vision of population is also produced through QB, which has curated samples in an order that recreates a "normal" demographic distribution by foregrounding the role of Qatari men and women with a relative balance in the demographic makeup of the biobank. The biobank, therefore, achieves several things. It creates a material resource of biological samples that can be recruited to draw human capital to Qatar in the form of scientists, clinicians, nurses, and health-care professionals who connect Qatar with the global infrastructure. QB achieves a Qatarization of the samples in the biobank, rendering Qataris a privileged cohort, which in turn helps promotes the imagination and performance of the Qatari nation as a lived reality.

5 WHOSE DNA IS IT, ANYWAY?

Khalid Fakhro strides from his Land Cruiser toward my table at Layali Aley. He picked the Lebanese restaurant as it is not far from his lab at Sidra and because he heard it has a reputation for good Levantine cuisine. When I'd last met Khalid for dinner two years prior, he wore jeans and a T-shirt, but today he's dressed in the Khaliji snow white thobe, topped with a draping crimson keffiyeh. After he greets the waiter in Arabic, we are ushered upstairs to a table on the balcony. It's December, and the late afternoon breeze is pleasantly mild.

I've looked forward to Khalid's presentations on the Qatar Genome Programme at each of the genomics symposia I've attended in Doha. Khalid runs a research laboratory at Sidra, and his team recently published the first Qatar reference genome (a map of the genetic variants in the Qatari population). His talk at Sidra a few days prior to this meeting had revealed new insights on the mixed genetic origins of the Qatari population. We order a spread of fattoush, grilled kebab, stuffed peppers, and the house special, a truffle hummus. The food arrives in installments without interrupting our conversation. Khalid shares a little more of his biography in accentless English as we begin sampling the food. He grew up in Bahrain but went to college at the University of Chicago and did his PhD in human genetics at Yale. He explained that his timing was fortunate:

> It coincided with the creation of next-generation sequencing technology tools, and them rolling it into academic environments. I happened to be working on some diseases in a big lab, and that lab was very interested in

> adopting next-generation sequencing technologies, so we were one of the first labs in the world to start doing that on disease families, and, by virtue of being at the right place at the right time, I started interacting with next-generation sequencing data on families. So that's how I started. I started by looking at Mendelian diseases. Specifically, these rare diseases that presumably have single gene mutation origins. Then, as a consequence of moving to this part of the world, where basically no genetics were established at all, the original goal was to come here and set up a Mendelian program for this population, but what ended up happening was that we started getting more and more samples, and then I got into more and more into population-genetics stuff.[1]

Khalid tells me he's now been in Qatar for eight years and explains how genetics research on the Qatari population has progressed. As he had mentioned, he initially began by working on rare diseases in the Qatari population, looking at Mendelian disease, rare diseases that presumably have single-gene mutation origins. As the technologies have become more efficient and cheaper, and as the Qatar Genome Programme was launched, the genomic sequencing of the population has sped up. We chat about his time in Qatar, and he explains how his own scientific biography is intertwined with the development of the Qatar Genome Programme. One of the topics we talk about at length is the idea of an ethnic reference genome.

THE QATAR GENOME AND THE ETHNIC REFERENCE

Determining the "genetic structure" of the Qatari population is one of Sidra's "Grand challenges" (Sidra 2014, 11). This determination involves an assessment of disease prevalence or predisposition and accurate phenotyping of the Qatari population. The strategic plan document sends a mixed message about the genetic nature of the Qatari population: it states that "Qatar has probably the most diverse patient population in the world" (37), even while it is somewhat unique because of its history of endogamy. Regardless of this ambiguity "many Qataris share recent common ancestry and also have large families" (47), making it easier to identify the underlying basis of genetic disorders that present as fetal anomalies. The Qatari population has a high consanguinity rate (54%) that brings with it an elevated incidence of recessive genetic disorders (47). Consequently, Sidra has been establishing a center for genetic and genomic medicine. Part of this project is a plan to "develop a prenatal whole genome sequencing research program" (48).

The goal of this program is to reduce the burden of childhood disease associated with autosomal recessive single-gene disorders. One of the most ambitious aims of the plan is to "do whole genome sequencing (WGS) of 10,000 Qataris (3% of the Qatar Genome project)" (Sidra 2014, 52). This target created the largest international genome project of its time when it launched, comparable to the ongoing sequencing project of the Genomics England project[2] or Singapore-based GenomeAsia100K (McGonigle and Schuster 2019) in terms of the amount of data generated.

The large-scale genomic sequencing of the Qatar Genome Programme has yielded a high-resolution characterization of the Qatari genome structure, which Khalid and colleagues published in 2016 (Fakhro et al. 2016). This is the first Qatar reference genome, which is being used to develop robust neonatal screening and assessment of genetic disorders prevalent in Qatar. The Qatar reference genome is a map of both the rare and the common variants that are present in the Qatari population. This reference genome may also be used as a model for the study of complex diseases in other populations. Khalid tells me that one of the notable successes of the Qatar Genome Programme and Weill Cornell Medicine–Qatar is the development of the first population-specific screening array—the Q-chip—which could be used instead of both newborn and premarital screening programs to determine risk for recessive disorders known to arise in the population. The Q-chip contains the gene variants specific to the Qatari population so that clinical diagnosis of genetic diseases using the Q-chip will be based on the unique genetic information derived from the Qatari population (Qatar Genome 2018).

During my meal with Khalid I ask him about the consequences for how Qataris understand ethnic and national identity in relation to the Qatar Genome Programme, and he explains that the research on Qataris has revealed a mixed ethnic heritage of the population. Initial efforts were to genotype the Qatari population based on the ancestry of the individual. He says,

> the International Hapmap Consortium had all that stuff and had already defined what European was, what African was, what Asian was, what East Asian was, what South American was, etc. We could tell that we had people who looked like they had ancestry similar to Europeans, or ancestry similar to Persians, but they are still very vague.

It is this vagueness that intrigues me. I wanted to know whether it is clearly written in your DNA which ancestry group you belong to or whether the grouping is dependent on known information about your ancestry. Other work from Weill Cornell Medicine–Qatar has established that the Qatari population has Bedouin, Persian–South Asian, and African components (Rodriguez-Flores et al. 2014).

This study by Rodriguez-Flores and colleagues is related to the Qatar Genome Programme, an affiliated project that uses genomic data provided by QB to examine disease prevalence and prevention in the Qatari population. Work from the Qatar Genome Programme has featured prominently in each of the annual functional genomics symposia I attended. The Qatar Genome Programme too is cognizant of both the biological and social factors behind genetic disease. For example, the aim of Sidra's 2016 annual conference, titled "Functional Genomics and Beyond: Nature via Nurture," was to address "one of the oldest and most challenging questions in science" (Sidra 2016). The symposium brought scientists and clinicians from around the globe to present the latest discoveries on the interactions between nature and nurture, and how these complex interactions are implemented in health-care practice (Sidra 2016). Their use of the idioms of nature and culture reveal an acute awareness in Qatar and in the broader Gulf region that the origins of genetic disease are a consequence of spontaneous genetic mutations whose effects are exacerbated by marriage practices and large families that increase the risk of disease propagation.

One of the speakers at this symposium was Myles Axton, the chief editor of the journal *Nature Genetics* (which sponsored the symposium), who also spoke at the 2015 meeting, when he urged the Qatar Genome Programme to publish their data in a public database. During his 2016 lecture, he mentioned the impressive work on precision medicine emerging from Saudi Arabia, one of Qatar's rivals in the field of genomics. Axton specifically noted the current issue of the top scientific periodical *Science*, which published an article on the clinical application of functional genomics in relation to inheritable disorders in Riyadh (Kaiser 2016). In a similar vein to the symposium theme, the article was titled "When DNA and Culture Clash." Accordingly, the cover of the issue featured the title "Family Ties: Saudi Arabia Strives to Prevent Genetic Disorders," with a picture depicting a Saudi family facing away from the camera and walking toward the sea (figure 5.1).

The *Science* article profiles a young Saudi clinical geneticist, Fowzan Alkuraya, who had recently returned to Saudi Arabia after working in the United States, and who had also given a lecture at the Sidra 2015 and 2016 symposia that I had attended. The author explains how genetic disease in Saudi Arabia results from an entanglement of kinship practices and genetic determinism or, in the editorialized glossing, a clash of DNA and culture. Strong adherence to tradition "helps explain why about 40% or more of native Saudis—two-thirds of the country's 30 million people—still marry first cousins or other close relatives . . . helping preserve wealth and tribal ties" (Kaiser 2016, 1218). The cost of this tradition of consanguineous marriage, Kaiser reports, "is a relatively high risk for recessive genetic diseases, which develop when both the maternal and paternal copy of a gene are faulty" (1219).

In Saudi Arabia, where large families are still common, "the genetic dice are rolled repeatedly" and "by one estimate, 8% of babies in Saudi Arabia are born with a genetic or partly genetic disease, compared with 5% in most high-income countries" (Kaiser 2016, 1219). The same article includes a short insert on developments in genetic medicine in Qatar, titled "Qatar's Genome Effort Slowly Gears Up." It reports that there have been difficulties involving data sharing and privacy in Qatar:

> So far, no outside researchers have gotten their hands on the information, as QGP [Qatar Genome Programme] officials and scientists wrestle over data access issues. These issues include how to prevent the DNA sequences from being downloaded onto other computers or accessed from outside Qatar, and who should be liable if people's genetic or clinical information gets stolen. (Kaiser 2016, 1220)

New books for the kid scientist's library *p. 1222*

Expanding financial services in Kenya *p. 1288*

Designer cells to correct insulin deficiency *p. 1296*

Science

$15
9 DECEMBER 2016
sciencemag.org

AAAS

FAMILY TIES

Saudi Arabia strives to prevent genetic diseases *p. 1217*

Figure 5.1

Family ties: Saudi Arabia strives to prevent genetic diseases. Logo and cover text from the December 9, 2016, cover of *Science*. Reprinted with permission from AAAS. Background image used under license from Reuters, © Adrees Latif/Reuters.

My conversations with senior leaders at Sidra confirmed these challenges, which admittedly have made it difficult for smooth collaboration between Sidra, the Qatar Biobank, the Qatar Genome Programme, and other stakeholders. In particular, several parties disagree over who should be legally responsible for protecting the human genomic data they jointly generate. This perceived risk is heightened by the liability that nobody wants to assume should there be a data breach that exposes the disease risk of identifiable and influential Qataris. Regardless of this hurdle, perhaps one of the bigger challenges is the political sensitivity of balancing the message of Qatar being a population of mixed origins. A message of mixed origins could potentially stand in contradiction with the state-led effort to emphasize the unity and independence of the nation while facing regional political isolation. Let us explore the details of the genetic origins of the Qatari population to better understand the issues of national and tribal identity in the region.

MIXED ORIGINS

The Qatar Genome Programme uses exome sequencing—coding DNA that determines a specific phenotype—in order to identify risk variants for Mendelian disorders that are prevalent in Qatar (Weill Cornell Medicine–Qatar 2013). They have already sequenced the DNA of 15,000 samples, and as stated above, this work has revealed three major ethnic subgroups of the Qatari population: people of Bedouin, Persian–South Asian, and African descent. By tracking the Y-chromosome lineage of the population—that is, tracking by patrilineal descent—the population is understood to have Bedouin (Q1) (Rodriguez-Flores et al. 2014), Persian–South Asian (Q2), and African (Q3) components (see figure 5.2).

Over our meal, I ask Khalid to explain about how the Q1, Q2, Q3 rubric was formed, and what it means in terms of defining an individual person in terms of ancestral heritage. He replied:

> It's iterative. The more people you sequence, the more ethnic population, the more labels you put on your map, the better you start defining what you want. Instead of sequencing, you say, you know you're 30% Irish and 70% German; I would say, oh you know, you're like, I don't know what are the

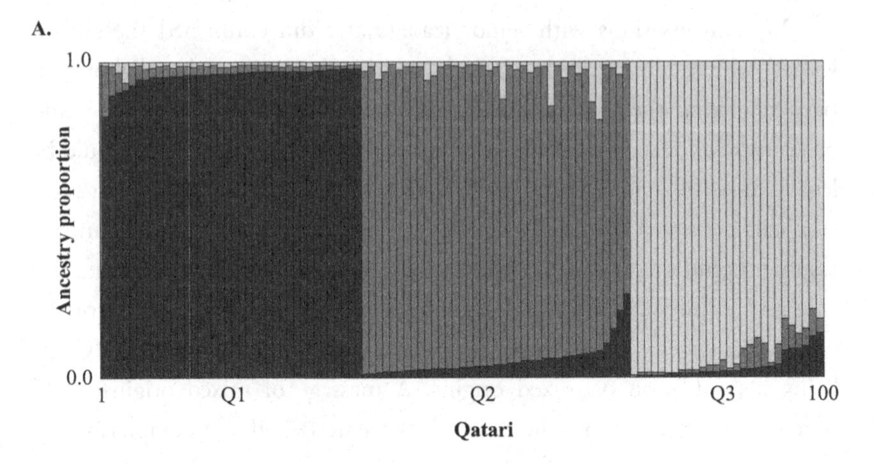

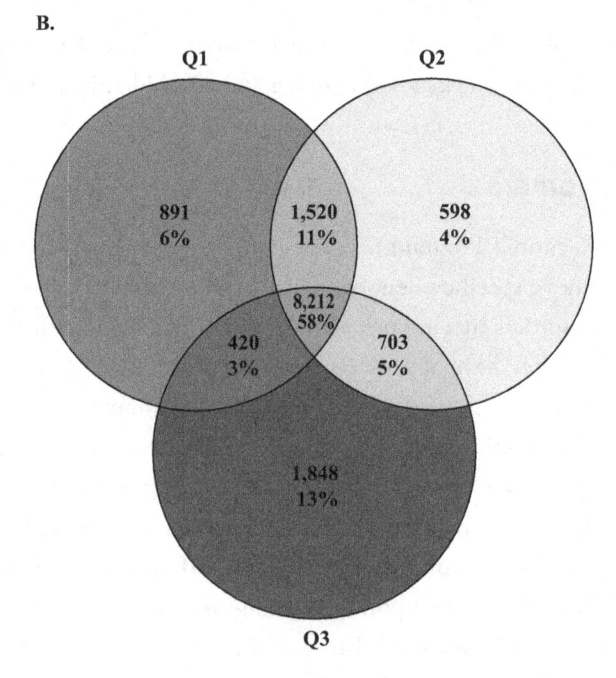

Figure 5.2

The genetic structure of the Qatari people. *A*: Qatari subpopulation structure. *B*: A Venn diagram showing the overlaps among three sets of Bedouin (Q1), Persian–South Asian (Q2), and African (Q3) subpopulations. From Rodriguez-Flores et al. (2014), with permission from John Wiley & Sons.

tribes inside Ireland, but I'll say, like, 10% this tribe, 10% from Dublin, or Belfast, or whatever.

I am pleased to hear Khalid's thoughtful response, one that reveals a skepticism about the natural facts of identity and instead emphasizes the heuristic, pragmatic, and contingent character of identity in the service of genomics. If it is more analytically useful to use tribal identities over national identities, if tribal identities can reveal more about the distribution of rare genetic variants, then it would make more clinical sense to move in that direction. I respond to his comment: "Right, you fraction it. Say you take the Persian component of the Qatari population; in itself it's a composite of many other different migrations and so of course these reference populations are essential to come up with the allele frequencies for diseases in a particular group, but there's a limit to their clinical utility?"

Khalid catches my line of thought on the flexibility of ethnic labels, and he explains further, "So the idea that you belong to Q1 is not mutually exclusive with you having Q2 haplotypes in your genome. Because at some point, one of your ancestors might have mated with a Q2." He makes it clear that, in genomics research, the attribution of an ethnic identifier does not say anything about the individual's contemporary social identity, nor does it preclude mixed ethnic heritage. Rather, it is a rough assessment of genetic ancestry for the purposes of tracking the distribution of genetic variants. But these rubrics are not fixed, and Khalid explains how, even more recently, the three Q components are being further divided as more Qataris have their genome sequenced:

> Now, more recently it is shown that Q1, Q2, Q3 is naïve, and we now think there are five Qs, that Q1 can be further subdivided into Q1a, Q1b, and Q2 can be further subdivided into Q2a and Q2b. So the one that was Persian South Asian, now you can see the South Asian and Persian split, and this is really the product of sequencing more people, with 100 people it was Q1, Q2, Q3. Now that we have 1,500 people sequenced, or as the QGP guys have, 6,000 people, 10,000 people sequenced.

As time passes and as sequencing becomes cheaper and more widely available, the ethnic stratification may continue to iteratively divide and include more and more identities or even artificial subgroups. But where is the limit? Could we reach a point where ethnic genetics is no longer the paradigm at all?

YOU ARE YOUR OWN REFERENCE GENOME

"At the end of the day the idea is that we are each our own reference genome," Khalid tells me in response to my questions about the limits of the utility of racial or ethnic identifiers in genomic research. I recognize nested in his response a way out of the difficult social and political worries that hang over the genomics of ethnic populations. I also hear the utopian vision of precision medicine. One day, in the future, we might all have our full genome sequence on demand, and it could be the clinical reference for our genetic disease risk. This is the ideal in the current thinking on precision medicine.

In this view, the ethnic identity of the components of the Qatari population is not assumed to be of great biomedical significance in itself, but understanding the genetic diversity of the Qatari population is essential if disease markers are to be distinguished from the natural genetic diversity within the population as a consequence of its origin history. Meanwhile, however, a Qatari reference genome is critical in the effort to identify the genetic bases of disease in the Qatari population.

So far, by analyzing Qatari genomes and correlating the data with the online databases of the 1000 Genomes Project and other resources, the researchers have identified the common variants that relate to disease in the Qatari population. One reason for the high rate of genetic disease they are uncovering is the historically high rate of consanguineous marriages, which increases the frequency of monogenic diseases appearing. But it is important not to single out Gulf populations and stigmatize them. Dr. Ronald Crystal, of the Qatar Genome Programme, said:

> Disorders are present in all populations around the world, so it's not the case that Qatar is different. Qatar is only different in that its variations and the frequency with which they occur are unique to its population. By finding out what these variations are and taking appropriate action we can save people

from the trauma of some very unpleasant disorders. We're talking here about things like brain malformation, diabetes, blindness, deafness, cardiovascular disorders, inflammatory disorders and many other conditions. While these conditions are not common, they do occur, some are untreatable and many are very difficult to live with, for both the sufferer and their families. (Qatar Foundation 2014)

After identifying the disease markers, researchers like Crystal hope it will be possible to eradicate inheritable diseases through premarital counseling and genetic screening. Prospective parents could undergo screening to see whether they carry the genetic variations that cause disease. The individuals that carry genes for the disorder would not necessarily have the disease, but they could carry and transmit the recessive genes to their offspring. Premarital counseling in Qatar initially screened for only four genetic variants, but the study found thirty-seven variants. Crystal further explained the possible practical applications of the findings:

> With more comprehensive screening, people will be able to make more informed choices about whether they feel it's safe to have children together. . . . Alternatively, it is possible to screen the fertilized eggs for variations that cause disorders before they are implanted. The improved screening can also be useful for adults who can change their lifestyle to prevent themselves from developing diseases. For example, if I analyze your DNA and tell you you're susceptible to having elevated accumulation of lipids—cholesterol and triglycerides that can cause cardiovascular disease—then you could alter your diet and take care to take plenty of exercise to mitigate the risk. (Qatar Foundation 2014)

In order for scientists and clinicians to reach a point where such robust statistical associations can be made, DNA and medical data from a significant section of the population need to be correlated and analyzed effectively. Collecting human samples and sensitive medical data raises ethical questions.[3] But genomics may pose challenges beyond medical ethics. Genomics may reconfigure understandings of ethnic origins, and genomics may present a challenge to religious understandings of the self, the body, social obligations, and individual rights in that regard.

So far, I have looked for indications of the sort of moral communities QB and the Qatar Genome Programme are attempting to build. But I also gained a some understanding of how local actors have tackled moral and ethical questions by considering communications among QB, the Qatar Genome Programme, and Islamic religious authorities. Indeed, QB and the Qatar Genome Programme raise questions that need to be squared with long-standing discussions in Islamic ethics.

"YOU ARE NOT THE OWNER OF YOUR BODY"

I first came across Mohamed Ghaly during my first trip to Doha in 2015. His lecture stood out among those on biology and medicine at Sidra's 2015 functional genomics symposium because the topic was Islamic ethics in relation to the Qatar Genome Programme. Having spent time in Israel and having become familiar with the ways in which Jewish religious law plays out so complexly in relation to the production of Jewish identity (see chapter 2), I was interested to learn from Mohamed about the Islamic mode of reasoning in relation to genomics. I wanted to find out how Islamic ethics may have something to say about the Qatari national identity and its potential authentication or demythologization by genomic analysis. I wanted to know what Islam had to say about the ethics of measuring the genetic origins of an ethnic population.

Leaning back in his office chair, Mohamed chuckles and tells me that when QB and the Qatar Genome Programme initially approached him to conduct an Islamic ethics assessment of their sampling practices and sequencing plans, the institutions' leaders thought this would be a quick and easy job. It turns out that Islamic legal scholars have a long history of thinking about the ethics of the body, and sophisticated guidelines already exist for a range of relevant problems. It's hugely complex, he tells me. For example, "When it comes to organ transplantation, . . . you cannot sell your body. . . . Even in the case of necessity." Your body is not yours for sale in Islam. The parts of the body that are separated, however, "like milk, human milk, you can sell, according to some religious scholars." It immediately becomes clear that such legal precedents would make the operations of a national biobank a legal and ethical minefield. What about blood, I wonder? Can you sell blood? It's technically an organ, but in some ways its partibility makes it closer to milk. Or DNA—do I own my DNA, and can I sell it? Or my genetic data, is that also my body? My mind races as I consider the myriad implications for a national biobank and genome project. These questions keep Mohamed very busy, as there are already hundreds of years of Islamic legal responses he needs to consider for guidance.

Islamic ethics in relation to genomics is a highly complicated field of thought, but one in which Mohamed has carved out a professional niche as a global leader. He is professor of Islam and biomedical ethics at the Center for Islamic Legislation & Ethics at Hamad Bin Khalifa University, Doha. When I meet with him later in 2018 to catch up and hear his thoughts about recent developments, he is mostly concerned about the growing valuation of genomics and big data and the entailed problems of privacy, data protection, and benefit sharing. He says:

> One of the things which I think still needs discussion is the ownership of the DNA. I, as Mohamed, went to you, Qatar Biobank, and I gave you the sample. You did research on this, you got funding through multinational companies, whatever, or from the Qatari government; you published the data, the data was taken over by pharmacogenomics companies, they developed a drug, . . . this drug will bring hundreds of millions. Who has the right to own this stuff? The general trend we have now, it's the [company] who catches the money. . . . But this company can never, could never get to this money without what happened before.[4]

The danger is that companies downstream of biobanking will capitalize on individuals' genomic data and that the donors would be cut off from the value that they have contributed, gaining nothing from the profits they helped generate. Mohamed thinks that if people understood the value they are helping generate, then "90% of the people will say no, I want part of this money."

He elaborates, explaining how this is not simply a question of social justice or the appropriate distribution of proceeds. The issue runs much deeper—to issues in Islamic law and an understanding of the individual self as the custodian of a body that is on loan from God.

> It will make a problem. . . . Maybe it's not a legal one, because you have, okay, you have informed consent, you are protected. But you will have a social problem, you will have an ethical problem, even ethical scandal, and I can say from an Islamic perspective you will have a very difficult question. It needs deep research and deep excavation. Because in Islam, you are not the owner of your body. It's not like secular bioethics.

In the light of these pressing Islamic concerns, QB and the Qatar Genome Programme made a fatwa[5] request to determine the moral, ethical, and legal implications of the biobank and its practices. QB leadership asked a series of questions. For example: is it "permissible in Islam to store biological samples for a long time after the death of the participants?"; is it "required to get the consent of the participants?"; does taking a blood sample nullify fasting; and is "donating samples an act of goodness and can it be part of Zakat and Sadaqa?"

They also asked for a determination of whether the biobank is in keeping with Sharia law and asked, "What does Islamic Sharia law say about fees to use preserved human samples" (Qatar Biobank n.d.-d). The fatwa reported, "Taking biological samples is permitted in Islam as long as their purpose is to serve research and studies that will benefit society and humanity." The fatwa also declared that "it is permissible to take, store and make use of these samples for research, whether during the life of the participant or after their death." However, the fatwa stipulated, "it is required to get the approval of the donor before taking the samples, and explain that they may be used in scientific research in their lifetime or after their death." It demanded that "the participant should be clearly briefed about the purpose for which the samples are to be used," emphasizing that "this right is permanently exercised by the participant, even after the samples are taken from them, and they may withdraw their approval or require that they must not be used after their death" (Qatar Biobank n.d.-d).

Most of the recommendations are in accordance with global normative standards, but they also are instructive for an Islamic way of life. The fatwa also advised people to participate "as long as the purposes of this research center in utilizing the participant's samples are useful and scientific" (Qatar Biobank n.d.-d). The fatwa concluded, "Most Islamic Madhhabs (schools or doctrines of law) agree that drawing blood samples do not [*sic*] impact fasting." As to whether giving samples is "part of Zakat and Sadaqa," the fatwa stated that "giving samples supports useful scientific research and is categorized as an invaluable deed of goodness and acts of righteousness and piety," but stated that "it can't be described as charity or Zakat because there is no possession involved, as the samples are not owned by the research centers or researchers" (Qatar Biobank n.d.-d). The issue of ownership comes up again and again, as in Islam one is not the owner of one's body in the sense of private property.

Since in Islam the body is not considered the sovereign property of the individual, alienation of bodily substance cannot be considered commensurate to a gift of charity. Normative ideas of property often follow a Euro-American construct that actually comprises a bundle of distinct rights, including a naturalization of private property; rights to exclusive use of property; rights to rent, sell, or dispose of property; and rights to alienate property by gift or payment. Such normative standards of rights presuppose a much wider set of economic social practices. In Islam, however, none of these assumptions can be taken for granted and the issues pertaining to genomics and biobanking must be considered in depth.

In thinking about the progressive valuation of data, particularly genomic data, I asked Mohamed whether the data derived from measuring your DNA are also considered your body and subject to the same Islamic regulations preventing their sale. He smiles and says, "This is another level of the question. The issue is that the data coming out of your body cannot be done without your body, but cannot be done with your body alone." It would seem that in the Islamic mode of reasoning, genomic data pose a conceptual challenge that Mohamed enjoys grappling with. In light of the complexity of these issues, the fatwa also addressed the moral value of charity in relation to biobanking and genomics.

Even if the sale of genomic data could be an Islamic ethical problem, the fatwa made it clear that the biobank and genome programme may offer the chance for performing good deeds. It stated, "Giving samples for research purposes can be regarded as charity as long as the samples or data are available for scientific research," and the fatwa stated that "the donor will be rewarded if they have good intentions" (Qatar Biobank n.d.-d). In this regard the fatwa recognizes the good deed of helping biomedical research, but the individual's intention is key. The fatwa also advanced an opinion about the guiding principles that the biobank had established. Such guidelines may be considered similar to the ethical norms governing the protocols of an institutional review board in the United States or elsewhere. In accordance with the fatwa, the biobank's guiding principles governing the research include the following:

Biobanking should cause no harm.
Biobanking should stick to transparency, the participant should know the purposes and uses of their samples.
Biobanking should keep privacy and personal information discreet according to the preference of the participant.
Biobanking should use the samples in research directly.
The samples must be used for research that benefits society.
The participant retains the right to withdraw their approval any time regarding the storage and gathering of samples and data. (Qatar Biobank n.d.-d)

These principles, of course, are not de novo formulations but rather echo long-standing values from Islamic law. The fatwa concluded that

> these principles are aligned with Islamic Sharia law such as "No harm for oneself and no harm for others, for seeking knowledge," is a condition set by the particiapnt [*sic*]. These are all principles applied by Islamic Sharia law that adheres to every principle that applies to the general morality in common between all people. (Qatar Biobank n.d.-d)

In relation to the issue of the ethics of cost and funding to support biobanking research, the biobank also requested an opinion on what Islamic Sharia law says "about fees to use preserved human samples." The fatwa stated,

> Imposing fees on donations is not accepted in Islam as the bank does not have the right to sell them as these samples can't be owned. But if these fees are for administrative work such as sorting, storing, and other services that have nothing to do with possession or purchasing samples, Qatar Biobank is permitted

to charge some fees for these services, but they can't be prohibitive, and can only cover the cost of administrative work. (Qatar Biobank n.d.-d)

Perhaps one of the most challenging issues for Islamic ethics in relation to genomics research is the management of incidental findings. Incidental findings are "results that arise although they were not part of the original purpose of the research project or clinical test." Such findings can be ethically problematic because "they can be lifesaving, they can also lead to harmful consequences for the individual and community at large" (Ghaly et al. 2016, 4). For example, finding a pernicious genetic marker in a certain individual, family, or community could spread stigma and make it difficult for individuals to find marriage partners. Further, knowing that one carries a disease gene could have detrimental consequences on an individual's mental health and well-being.

The issues are not just of concern to the individual, however. Finding genetic markers that are shared among a wide group, like the nation, may also impinge on the limits of belonging and the imagination of relatedness. Although genetics research on ethnic populations in the service of promoting health is welcomed, work that would split the Muslim nation into subgroups based on genetic data would be problematic from the perspective of Islamic ethics. It would challenge the unity of the nation, in this case, the nation of Islam rather than the Qatari nation, which could be splintered into subgroups. That, in principle, is un-Islamic. It is thus important to understand how genomic analysis of the ethnic Qatari population may challenge or assert assumptions about identity and belonging in the broader Muslim-majority region. This brings us back to the question of what is national about the Qatar Genome Programme, as well as the related question of how the Qatari genomic citizen should be understood in relation to Qatar's neighbor states.

A NATION AMONG THE TRIBES

When it comes to marriage practices, tribal boundaries have been respected for a long time in the Arabian Peninsula. This means that inheritable diseases in the region are likely to be associated with identifiable tribal lineages, making the long tribal history of the region an important factor in the genesis of rare genetic disorders. As Khalid tells me, "Usually, you know, families know that they have these things. Because they have seen them more than once."

The scientists are aware of the way in which genomics may link disease histories with specific family lines. Saudi clinical geneticist Fowzan Alkuraya presented a slide at Sidra's 2016 functional genomics symposium that depicts the original tribal identities of the Arabian Peninsula (figure 5.3).

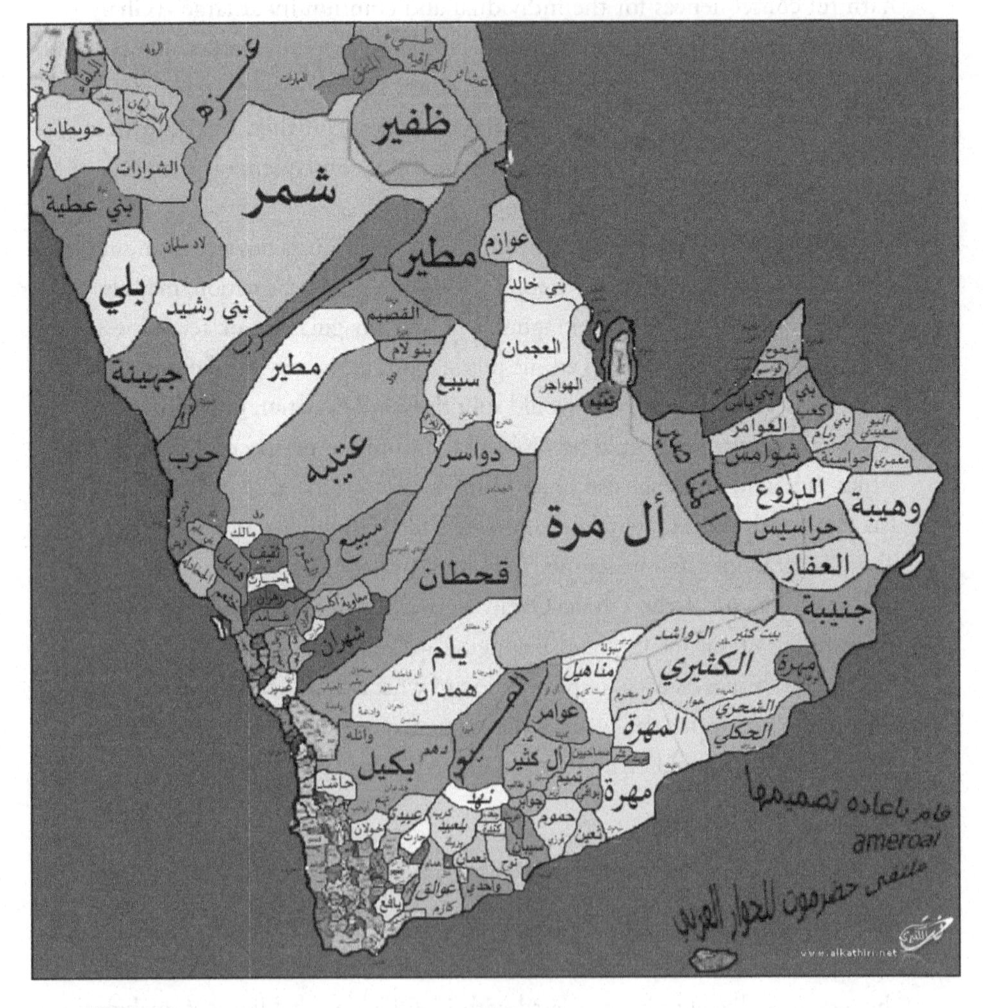

Figure 5.3

Map of the tribes of the Arabian Peninsula. © Emirates / Wikimedia Commons / CC-BY-SA-3.0.

The implication is that genome projects that unearth deep tribal history in the Arabian Peninsula could impact contemporary tribal identities, but in what manner and how likely this is remains unclear. For example, if certain diseases or benign biomarkers are identified and associated with specific tribal groups, then tribal identities will be able to draw on the genomic data as a verification of their authentic identity. More specifically, if a disease becomes publicly associated with a family name this could create devastating stigma that could ruin a family's reputation. On the other hand, incidental findings from genomic studies could show a high degree of admixture, evidence that could potentially challenge the oral narratives of endogamous groups. As noted in the preface, the imagination of the Lebanese people as being genetically distinct from their Middle Eastern neighbors has begun to enter mainstream political discourse.

The science could be appropriated as a rhetorical device in the service of a multitude of constructive or destructive sociopolitical projects, be they tribal alliances, national loyalties, or claims of ethnic superiority. But the contemporary formal political organization of the Arabian Peninsula in nation-states shapes the way in which the genomics of ethnic populations takes form. In a national context, genome projects emphasize a relatively novel national character of the populations. The nation is assumed to be the overarching category of analysis, even though populations have been mixing for centuries in the region. One explanation for this assumption is simply the pragmatic nature of state-driven biomedical developments, but that contingency has important consequences for ethnic identity.

Genomics affords, indeed invites, the imagination of the nation as a real thing. Philosopher Slavoj Žižek (1993) equates the imagination of the nation with transference—or redirection—of the desire to reconnect with the vanishing symbolic "other" via the mediated sign of the nation. The nation-state, in his reading, which he calls the "Big Other," serves as an object of desire that one seeks to identify with. In reading nation as fiction, he claims, "'Nation' is a fantasy which fills out the void of the vanishing mediator," providing an "idea thing" that can be performed or enjoyed as a "Nation-Thing"—a national essence to be "grasped" and "enjoyed" as one's own (1993, 232). We should note that genomics is emerging in Qatar at the same moment that a national identity is being established. This context of nationalism thus conditions the field of genomics to inculcate authoritative fictions about the Qatari nation as a natural object that bolster the political establishment's nation-building initiatives.

The imagination of a shared history in the new Middle Eastern states, as elsewhere, is not uncommon. It is usual for "new" memories of "ancient" historical events to be recruited for the purpose of nation-building. In Lebanon, for example, the myth of the ancient Mediterranean seafarers and traders, the Phoenicians, held great ideological traction in the minds of the Christian Maronites, who saw Lebanon as a natural Christian country. With the rise of nationalism in the 1850s, the Christian Maronites were drawn to discussing the history of the Phoenicians, who they presumed to be their natural ancestors. Historian of Lebanon Kamal Salibi recounts how the "archaeological exploration of the Phoenician past of Lebanon, first by French then mainly by Christian Lebanese archaeologists was politically geared—officially as well as by private initiative—to strengthening the theory that modern Lebanon was none other than ancient Phoenicia resurrected" (Salibi 1988, 172).

Similarly, in the case of Israel, archaeology was once the "identity-forming practice par excellence" that could identify artifacts as "objects of history," actants that embody forgotten stories and lost memories (Feige 2009, 100). Science, with its implicit claims to disinterested objectivity, can thus play a key role in bolstering the legitimacy of historical claims. Of archaeology in Israel's early statehood years, Abu El-Haj writes, "through the very nature of archaeology's historical practice, epistemological commitments, and evidentiary terrain, it helped to realize an intrinsically Jewish space, continuously substantiating the land's own identity and purpose as having been and as *needing to be* the Jewish national home" (2001, 18; emphasis in original). She argues elsewhere (2012) that genomics is now the prestige science that holds particular rhetorical power in making historical claims in regard to peoplehood and their shared experience. With advances in DNA sequencing, population genomics has joined archaeology as one of the prestige sciences recruited for nation-building and for imagining a shared common origin. It is no longer only shards of ancient pots, sunken ships, or scraps of biblical texts that capture the minds of people with epistemic power and help them imagine history as an objective truth. Biology is becoming a more dominant method for defining authentic peoplehood while also gaining higher credibility.

The Qatar Genome Programme also gives citizens a chance to perform their nationality and citizenship by giving blood, a potent symbol of life, solidarity, and relatedness. Moreover, this act of donation is now proclaimed as a good deed by Islamic authorities in the form of a fatwa. Participation in QB and the Qatar Genome Programme allows Qatari citizens to imagine also participating in a national population in a way that complements the emerging culture of Qatari nationalism.

In chapter 4, we saw how Qatar's biomedical research targets Qataris to be included as high-skilled workers through policies of Qatarization. But Qataris are not only the *subjects* of such state education and employment initiatives. In QB and the Qatar Genome Programme, Qataris themselves become valued *objects* of biological study. In fact, Sidra's strategic plan is explicit about objectifying the Qatari population as a biological object:

> The uniqueness of the Qatari/Arab population provides a special strength for the study of population structures, functional annotations of large homozygosity regions and other investigations at the interface between genetics and biological function that can be best studied in endogamous populations. It should be noted that while other large population studies have focussed on quite homogeneous populations such as the inhabitants of Iceland, the structure of the Arab people is unique, diverse and complex with influences from historical migrations since the dawn of civilization. Therefore, the richness of genetic information obtainable by studying Qatari/Arab people is a major asset for the world. (Sidra 2014, 6)

The plan emphasizes the special features of the Qatari population as a biological object as well as the commonality that Qataris share with the greater Arab people. The Qatari population is both unique and similar to the broader Arab population. It claims both particularity and a dimension of universality that echoes an ideology of pan-Arabism and industrial development.

In Qatar, the state and its sponsored scientists do not appear to be intentionally constructing an engineered genetic narrative about the origins of the Qatari people. However, a genetic narrative could nonetheless emerge later on as an unplanned consequence, as happened with studies that emerged from the Israeli biobank. So far, in QB and the Qatar Genome Programme, we see an entanglement of specific medical needs of a population that are the result of rapid modernization and urbanization (diabetes and obesity in particular) with the general desire for the state to be at the highest level of development in terms of personalized medicine. The unique features of the Qatari Arab population are asserted, while the mixed origins of the Qatari nation are acknowledged. The Qatari nation is presented as exceptional in the sense of it being a genetic mixture that gains privileged medical attention amid a regional blockade and a thriving culture of state nationalism.

6 GENOMIC CITIZENSHIP: HYBRID ONTOLOGIES

Global science takes on national characteristics in specific locales. However, a close examination of the laboratory practices, genome projects, and biobanks in Israel and Qatar necessarily leads to the conclusion that there is no unequivocal genetic marker or signature that is a sine qua non of Arab Qatari or Jewish Israeli identity. No Jewish genes, no Qatari genes. A national context does not entail a national gene. In fact, an internalist analysis, or immanent critique, of the scientific discourses suggests that it is precisely the genetic diversity of these populations that renders them interesting and valuable for medical research.

Although there have been other studies of genetics and medicine in the Middle East (e.g., Birenbaum-Carmeli 2009; Burton 2018; Falk 2017; Inhorn 2015; Kirsh 2003; Kohler 2014), the way biology is beginning to reconfigure ethnicity in the contemporary Middle East has been less well explored. In particular, little attention has been paid to the social life of genetics in the Gulf states. Given the lack of scholarship on this topic, I was surprised when Kuwait announced that it would become the first country to demand that a DNA sample of every citizen be assembled in a database. To be a Kuwaiti citizen, "according to the Kuwaiti constitution, citizens must be able to prove that they or their forebears have lived in Kuwait since 1920" (Cook 2016). If this law were strictly applied, "about 10% of the Kuwaiti population are not citizens" (Cook 2016). Despite the fact that there is no legitimate scientific basis for "identifying" in the philosophical sense of the word (i.e., equating ethnic concept with substance) a national essence legible in DNA, genetic tests may still be deployed to determine biological connections, and those biologized kin relationships may become the basis for determining ties that achieve national belonging and rights to citizenship. Or, at least, the imagination of genetics' "truth-power" has infiltrated the discourse of state biopolitics and border control. In turn, this kind of biopolitical discourse only strengthens the metaphorical potential of DNA as the bearer of an essential ethnic identity.

These findings raise a contradiction: despite the science falling short of unequivocally authenticating the bio-nation, the imagination of genomic citizenship persists and circulates, even at the level of the state. The social life of genetics exceeds its epistemic bounds in the affirmation of norms. How has this happened?

"IT'S IN OUR DNA"

The phrase "it's in our DNA" has been ubiquitous in recent years: you can hear it in everyday conversation, in an ad for a Land Rover, or by CEOs to describe institutional culture. Though this circulation of genetics as a metaphor for internal essence and inclusion is very real, the scientific justification is often weak. It is, of course, the primordial relations of inclusion/exclusion that charge the genetic metaphor with its rhetorical power. This relationship of genetic imaginaries and national inclusion/exclusion can be understood as the coproduction of a national imaginary and a genetic discourse (Jasanoff 2004). This is not to say that the nation is less real than the science, or vice versa. Rather, the nation as a lived reality is concretized in institutions of power and governance, while the bio-nation as genetic cohort becomes the so-called metaphysical add-on that achieves credibility only within the conditions of state power that can back up the fiction of peoplehood with truthful force. Fundamentally, it is states that create the conditions for the imagination of genomic citizenship, even while it is the language of science that propagates the illusion.

We know from Anderson's (1983) canonical analysis of the mediation of national identity in Southeast Asia that newsprint media inculcates a shared imagination of citizens as copresent members of an imagined nation-state.

Appadurai (1990) extended this reading to think multidimensionally about the diverse regimes of cultural mediation in a delocalized, globalized, but interconnected, world. The Middle Eastern ethnonationalisms I have tracked here, however, gain a particular symbolic traction in the biosciences. Genetics becomes a regime of mediating the nation. In Israel and Qatar, it is the confluence of state development goals, global scientific ambitions, internal demographic arrangements, and specific citizen medical needs that, in concert, render the "bio-nation" an object to be apprehended in the biomedical sciences. In such an assemblage as the bio-nation, the citizen is apprehended as a therapeutic subject (as in Qatar Biobank [QB]) or is imagined as a "natural" genetic constituent of the bio-collective, as in popular discourses of genetic belonging and citizenship in Israel.

COMPARING BIOBANKS

In Qatar and Israel, diverse factors, both local and global, influence biological research at the level of national laboratories and national biobanks. The political, economic, legal, and historical contexts of each state plays a role, as do global trends in precision medicine. The symbolic significance of each biobank, in terms of its demographic composition, can be read as ordering populations. Biobanks arrange a set of taken-for-granted identities that are institutionalized and thus rendered more authentic in the process. But between Qatar and Israel there are also important differences in the relationships among state, biobank, and global science. The representative function and character of the two biobanks differ in ways that can be seen in an abbreviated form in table 6.1. Crucially, the apprehension of the nation as a biological object is arrived at with a different degree of proportionate influence by the states of Israel and Qatar.

In Qatar, the biobank disproportionately represents the Qatari minority. In this regard, it is a national representative space that fails to proportionately include the presence of residents other than the Qatari minority, those who in fact constitute the majority of the population. This renders the Qatari identity a privileged status when it comes to access to health care. As Comaroff and Comaroff write, "In as much as collective identity

Table 6.1
Summary comparison of Israeli and Qatari biobanks

	NLGIP	Qatar Biobank
Initial funding	Israeli Academy of Sciences and Humanities (funded by the state)	The Qatar Foundation (funded by the state)
Genetic nature of the population	"An exceptional mix of varied populations from diverse ethnic backgrounds" (Gurwitz, Kimchi, and Bonne-Tamir 2003, 3)	"Qatar has probably the most diverse patient population in the world" (Sidra 2015b, 47)
Number of samples (as of November 29, 2015)	1,673 (unrelated)	3,022 (participants)
Current use	Becoming underutilized	Emerging and imagined to improve the health of the Qatari population
Genomic data	No	Yes
Future aims	Currently planning a transition to research on personalized/precision medicine	60,000 participants
Social contract	"National repository for human cell lines and DNA samples representing the large variation of Israeli and several Middle Eastern populations" (Gurwitz, Kimchi, and Bonne-Tamir 2003, 5)	"A scientific and altruistic partnership between the research community and the people of Qatar" (Qatar Biobank n.d.-a)
Biobank goal	Global biobanking	State- and capacity-building
Regulatory overseer	Tel Aviv University, Institutional Review Board (an academic institution)	Supreme Council of Health (a national ministry)

always entails some form of communal self-definition, it is invariably founded on a marked opposition between 'ourselves' and 'other/s'; identity, that is, is a *relation* inscribed in culture" (1992, 51; emphasis in original). Identity is as much about who is excluded as it is about who we are. Similarly, Dominguez, in her ethnography of Jewish identity as an historical "object" in Israel, claims that "the existence of 'objects' is always supported, challenged, bolstered, or molded by individuals and groups engaged in social and political processes of everyday discourse and institutional life" (1989, 21). Identity is thus often dependent on institutions for its legitimation and support.

These propositions demand reading ethnic identity as an inherently relational construct at some level. In the Gulf states, the rentier economy prefigures the value of citizenship in relation to migrant workers, whereas in Israel distinctions are made among citizens as to whether they are Arabs, Jews, Christians, Druze, and so on. Today, the National Laboratory for the Genetics of Israeli Populations (NLGIP), however, has a representative diversity of participants that roughly reflects the demographic character of the state's territory. It includes samples from the ethnic minorities that are not formally represented by the state, which is exclusively Jewish in character.[1]

The NLGIP itself does not explicitly emphasize the shared genetic basis of the whole Jewish people or the shared genetic heritage of the different Jewish ethnic groups, although some of the studies that have used samples issued by the biobank have made historical claims about Jewish origins. The NLGIP has not taken part in a state effort to fight a demographic war and diminish or exclude the Palestinian/Arab Israeli populations. Rather, the NLGIP was founded with the intention of serving a humanistic project of furthering biomedical knowledge of the global diversity of the human species. These goals are consistent with the discourse of development described by Shamir (2013) with the electrification of British Palestine and with early Zionist rhetoric of making the desert bloom. In Qatar, however, a new identity is being forged as QB is coproduced with a strong but emergent Qatari national identity. Indeed, QB and Sidra have stated intentions to uncover the genetic structure of the Qatari people, though it is unclear if this research will emphasize a historical presence in Qatar, justifying the sovereign rights of the state, or indeed if this "structure" will emphasize the diverse origins of the population shared with other Indian Ocean and Persian Gulf nations.

Important differences in terms of direct state support also distinguish the biobanks. The NLGIP does not receive direct funding from the state; it survives on individual grants that must be renewed, and the future of the biobank is uncertain. In Qatar, QB is in its early phase but has plans to grow significantly in the coming years, with generous funding from the Qatar Foundation, which is endowed by the emir. In this regard, the missions of QB and Sidra are the indirect implementation of the vision of the emir and his close family members. These biomedical developments are generating great activity and draw professional migrant workers from across the world. One reason for the interest and support that QB has engendered is its goal to be a top-level player in the global move toward personalized medicine, which requires complex personal, medical, and genetic information about large numbers of people. This fact ties the excitement about Sidra and Qatar to the global market rush in health care toward personalized therapeutic and diagnostic technologies.

The NLGIP, however, without extensive genomic data or sophisticated biomedical information, is becoming used less often in this current move toward big data analytics that foregrounds the search for biomarkers for precision therapies. The NLGIP must also be understood in the context of a state that has generally moved to the political right in the past forty years: the state has withdrawn from major public development in a context characterized by the emergence of a so-called start-up nation economy, characterized by the large numbers of entrepreneurs and successful innovative ventures it has produced (Senor and Singer 2009). Senor and Singer ask the "trillion-dollar question," "How is it that Israel—a country of 7.1 million people, only sixty years old, surrounded by enemies, in a constant state of war since its foundation, with no natural resources—produces more start-up companies than large, peaceful, and stable nations like Japan, China, Korea, Canada, and the United Kingdom?" (2009, i). Senor and Singer argue against reductionist and racist explanations and address the sociological factors that have conditioned the fabric of Israeli society and incubated this culture of entrepreneurship. They attribute Israel's success in producing entrepreneurs to several things. One is the loose hierarchical structure that Israelis learn in their mandatory military service. Israelis typically challenge authority and are rewarded for questioning authority if their objections are justified (Senor and Singer 2009, 23, 47); another is the high degree of technical training that is included in the special programs that the military provides. Moreover, the experience of military service leaves Israelis with a close network of friends and colleagues with whom ventures are typically launched.

Senor and Singer claim that "the IDF's [Israel Defense Forces] improvisational and antihierarchical culture follows Israelis into their start-ups and has shaped Israel's economy" (2009, 177). They also identify Zionism as an important imaginary in keeping Israelis motivated in helping build their nation. In comparison, they note the absence of a similar motive in the Gulf states of the Gulf Cooperation Council. They note that in Dubai "most of the entrepreneurs that come from elsewhere are motivated by profit—which is important—but they are not also motivated by building the fabric of community in Dubai" (2009, 205).

It would seem that a state's leading role in nurturing biomedical development, as seen in Qatar, is relatively absent in Israel, where private industry and entrepreneurship drive research and development and supplant the role of resources like the NLGIP in driving biomedical advances. In this reading, a strong national identity and mode of citizen identification with a nation-building project are necessary for prosperous growth.

In Qatar, there is a clear vision of capacity-building and attention to the citizen population's particular health needs. These developments are emerging directly from the epicenter of state power, through the Qatar Foundation, and embodied by Sidra and QB. In Israel, by distinction, the NLGIP exists and ensures its continuity in precarious circumstances, with no certainty and with no mandate invested in it by the state. Nor is it guaranteed future support. Nonetheless, both biobanks, and the associated research endeavors they engender, succeed in cultivating a vision of peoplehood, national coherence, and biomedical progress, which serve to foster an imagination of national community, meliorism, and a utopian future at the level of individual bodies and state economics.

TWO ETHNONATIONALISMS

Here we see two nationalisms at different moments in their development, but both coupled to genetics in complex ways. Israel has moved from its secular socialist beginnings in the early twentieth century to become a more globalized, commercialized, and divided society, with Ram's (2008) characterization of "McWorld" and "Jihad" (that is, globalized modernity versus its negation, local religious nationalism) becoming more pronounced. The contradiction between these cultural poles is apparent, for example, in the Israeli news media over the past years, oscillating from the forced deportation of African refugees to the latest start-up conferences and technical innovations emerging from Israel. Israel's character stands in tension between its dubious image as a state at perpetual war and its preeminence as a global innovation superstar.

The NLGIP may be understood within these poles of contradiction. The NLGIP represents the ambition of secular cosmopolitan Israel to be part of global modernity, and its work facilitates the "McWorld" integration and collaboration that is necessary to be at the forefront of biomedical research and innovation. The NLGIP stands for global modernity and secular technoscientific progress. It is within the other pole of the Israeli cultural spectrum that one finds right-wingers appropriating genetics as a way of imagining the tribal particularity of Jews, as a way of proving the occupation is legitimate, of authenticating the *ethnos* as a natural fact, and of defending Zionism as a return to primordial origins. It is across this political spectrum that the natural facts of genetics research discursively migrate and transform into the mythologized ethnonationalism of the bio-nation. However, Israel has also moved further toward a market-based society and the majority of biomedical research is moving to private biotech companies, leaving the NLGIP underutilized and outmoded. The epistemics of Jewish genetics fall short of its mythic circulatory semiotics. This is the ultimate lesson from my ethnographic work in Israel.

What is happening in Qatar is quite different. Qatar is in the early stages of its nationalism, and the modality of statecraft is distinct. Unlike Israel, Qatar is an ultra-rich monarchy endowed with an Islamic character. The Qatari state is massively investing in biomedical development as part of the national plan—known as Qatarization—to enskill Qatari citizens and include them in the high-tech economy. The state aims to make Qatar an equal partner with other developed states in global biomedical research. In terms of the circulation of national myths about genetics, Qatar also differs from Israel. In Qatar it is the state that is driving the genetic research of the Qatari people and selectively assembling Qataris for genomic analysis.

Nationalism in Qatar, however, is not part of the negation of a secular global modernity driven by the state. It is a way of building the Qatari national identity and driving technological development in concert. But the Qatar Genome Programme and QB need Islamic legitimization, particularly at the level of medical ethics, and this is a line of study that could be pursued further, with a careful look at how precision medicine gains an Islamic character in Qatar. However, there is not a widely circulating discourse of genetic nationalism in Qatar. I have not seen or heard debates about Qatari origins, nor has the state announced the use of genetics in determining citizenship. Perhaps the mythology of Qatari nationalism is amply filled by the state's symbolic efforts, with National Day, omnipresent flags, and a media image campaign carefully monitored by the state.

Regardless, it appears the Qatari national imaginary is bolstered by the genetics research concerning the Qatari people. The Qatari national imaginary is a structuring precondition of this relationship. The Qatari context does not square so well with Ram's (2008) dialectical formulation of secular globalism versus local tribalism. Rather, Qatar seems to be comfortably on the course toward global tribalism. Indeed, the Qatari state is ethnonational but with ambitions to be part of globalized modernity without negating the state's religious or ethnic character. The contradictions of "McWorld" and "Jihad" do not obtain in the same way in Qatar, where traditionalism and modernity sit alongside each other peacefully, in what Fromherz calls "neo-traditionalism," a blending of tribal traditions with a modern lifestyle, technology, and urbanization (2012, 113).

GENOMIC CITIZENSHIP

How can we better understand the ways the ethnonation becomes explicit content in the language of genetics? Earlier in this book I proposed that the process on which the discourse of the molecularization of ethnicity depends involves a double reification. That is, it depends on an assumption of unity between the concept of genetic signatures and the material reality of the ethnos, and that there exists a necessary reading frame, the felicity conditions of which are the doxic assumptions of the prior existence of an ethnos, that is, a people that exists with some shared characteristics and a belief in their sharedness. However, what I found demands a reconsideration of this proposition that a "double reification" would be simply an erroneous misrecognition of an existing ethnic essence that can be determined and legitimately read at the level of DNA.

Genomic citizenship is more than a misuse of science; it is a genre of imagining belonging. Set aside for now the meta-epistemological stance that would condemn the whole discursive arena of ethnic genetics to the realm of ideology and false consciousness. Instead, consider the precise mode of existence on which ethnic genetics depends. Or, more exactly, consider the ontological register on which the concept of ethnic genetics depends in order to come into existence. More precisely yet, consider the ontological foundation of ethnic genetics on its own terms, within the structure of its conceptual principles. In the most plain of language, the question remains: How is it that ethnic genetics has such a powerful circulatory potential—for example, in law, in the popular imagination, and in battles over historical presence—when at the level of molecular genetics there is an unequivocal consensus that there is no "ethnic gene," that the ethnos cannot be apprehended as a bounded, clearly defined entity?

The philosophical anthropologist Philippe Descola published a comparative study of the multifarious ways in which entities come into identifiable existence. He calls these distinctions "modes of identification," and they can be separated into four models that act as heuristic devices for separating entities that would ordinarily appear incommensurable. These models are ideal types. They are not supposed to represent the complex ways in which entities are understood or woven into daily practice. They are simply guides for comparing and analyzing the philosophical bases that underpin existence and modes of recognition in a cross-cultural context. The four "ontologies" that Descola proposes are an extension of long-standing anthropological concepts: animism, totemism, naturalism, and analogism (2013, 112). Descola is trying to consolidate the insights that the ethnographies of both modern and native societies have yielded over the past half century or so, and to formulate a way of thinking comparatively in a way that does justice to the diversity of ways in which humans establish social order and live in the world.

Naturalism denotes the dominant ontological scheme of modern secular societies, in which the evident physicality of what exists typically takes precedence over any invisible, internal, or immaterial properties of things. The naturalist ontology assumes we can understand the world based on the natural properties of entities. The distinction between internality and physical externality is a continuation of Cartesian dualism that punctuates the mind/body dyad and that is replicated in the soul/body distinction and nature/culture binary. Naturalism is the common sense of modern rationality, of secular societies, and of science. This is the world as we ordinarily encounter it.

Animism is the ontological scheme that accords with the proposition that what are typically deemed objects of "nature" (e.g., the sun, moon, plants, mountains, or rocks) may have internal vitalisms comparable with human interiority. In scientific communities, animistic beliefs have been impugned as primitive and unreasonable.

Totemism, which has been well cataloged in the ethnographic literature, is the mode of identification that typically denotes a relationship of equivalence, or identity, between a nonhuman entity and a human collective. An everyday example would be a team mascot that stands for the social collective and as such cannot be violated without implying an attack on the collective. The classical anthropological example, of course, is the clan totem (Durkheim 2008 [1912]; Lévi-Strauss 1971).

Descola's last ontology, analogism, is a synecdochic mode of identification that comprises heterogeneous elements and transcends scales, examples of which include horoscopes or traditional medical healing techniques that mobilize symbols for therapeutic efficacy. Analogism, in short, is a mode of identification that presumes a physical discontinuity, but a continuity of interiority between the objects in a relationship of identity.

Thinking across these ontological registers may be helpful in addressing the contradiction between the patent absence of an ethnic gene and the proliferation of gene talk, including the potential use of genetics as a regulatory technique of the state. For this discussion about the materiality/immateriality, and indeed the associated implications, of ethnic genetics, one might assume that the correct mode of identification for genetics, because it is a science, would be naturalism. In the naturalist ontology, a genetic marker for ethnicity would need to be physically evident by some means according to the accepted epistemological standards of the science.

However, even though there is indeed no unequivocal evidence of a "Jewish gene," or a "Qatari gene," the discourse of genetic identity still holds weight and achieves a certain amount of worlding. This is part of the basis for genetic research of national populations, and for the mythic circulation of "gene talk." Even though the ethnic gene does not exist by the internal naturalistic standards of genetics, it nonetheless exists and persists and may be better identified or apprehended in another ontological register.

Certainly, through the channels of discourse that it enables, ethnic genetics serves to stabilize the imagination of the ethnos. Or at least it may be mediated in a modality comparable with the standards of one or more of the ideal types that Descola has described. Even while not achieving the standards requisite for existence in a naturalist ontology, ethnic genetics still functions as popular discourse, a way of assuming, believing, and imagining, and as such it may still impact modes of managing national populations. It reiterates the ontic essence of the nation; it is also a way of measuring belonging (elective or imposed) of citizens. In an obvious sense, the gene talk emerging from ethnic genetics is patently bad science. The discourse can perhaps be understood in more precise anthropological detail by the ideal standards of the analogist ontology, or read as semiosis. Even while no material connectivity can be measured in genetics that could establish an externalized continuity between all the Jews or between all the Qataris, the imagination of some interior continuity is both assumed and propagated through the genetics of ethnic populations. The molecularization of ethnicity does not gain its epistemic power convincingly in a pure naturalist ontology. It is patently not evidenced. There it fails. Rather, the discourse of ethnic identity takes a hybrid form in the language of genetics, in which a relationship of collectivity is presupposed. This is the precondition, I submit, for why the imagination of genomic citizenship persists and circulates, even while it fails within a naturalist ontology.

The imagination of genomic citizenship is not a mediation of scientific facts; it is the dissemination of an identity that is the structuring frame within which the genetics research is made possible and conducted. This finding, the fact that the molecularization of ethnicity takes form in a hybrid mode of identification in genomic citizenship, has significance for the social study of science and especially for the study of identity. It demonstrates that in this era of the molecularization of identity, with the rapid proliferation of self-directed recreational genetic testing, identification is not sequestered and negotiated solely in a naturalist ontology. Even when identity cannot be verified in the natural sciences, the imagination that has propelled the effort succeeds. For this reason, one cannot adequately critique ethnic genetics without inadvertently critiquing the psychic foundations of the imagination of an ethnos.

The reiteration of commonsense assumptions about ethnic difference is not unusual in genetics research. Indeed, ethnic categories assumed within the Israeli population were reinforced in the work of geneticists on blood groups in the early state years. Abu El-Haj writes that "the classificatory categories 'Ashkenazi' and 'Sephardi' were black-boxed in the very design of the studies." Rather than questioning the basis for ethnic categorization, "those categories were assumed a priori to exist." Work at that time took for granted the natural facts of ethnic difference, and thus "reiterated the biological truth that the Ashkenazim and the Sephardim are identifiable populations" (2012, 93). Ethnic difference in early Israel was based on assumptions about distinctions between different immigrant groups even while those different Jewish groups were nested within an assumption about a joint common ancestry. Such a "genetics of belonging," what I am calling here genomic citizenship, can perhaps be imagined only in a context of belonging.

Ethnic genetics cannot be separated from nationalism. Consequently, genomic citizenship cannot be negated with the language of genetics. This would be a misunderstanding of the phenomenon. Rather, its ontological scheme is hybrid, both naturalist and analogist. It imports imaginaries that are not guaranteed by the science. This analogist mode of identification is constituted outside of a material, naturalist mode of identification. This proposition conforms with the insights furnished by the social study of science, which have demonstrated that science's epistemic outputs are never fully "purified" from their politicized context of production (Latour 1993). It is merely the modern conceit, or meta-narrative, that nature and culture are "naturally" separate domains. Hence, the nation can neither be proved nor disproved by genetics.

But each case of purification is unique. In Israel and Qatar, biobanking and genetics research relate to the ethnonational context and to their populations in varying but comparable ways. The molecular realm is clearly an emergent site for articulations of ethnonational identity in the contemporary Middle East in a multitude of ways: in national genome projects, biobanks, and in the entailed performance of biological citizenship.

These institutions and practices are entangled to varying degrees with popular understandings of emerging nationalisms, public health initiatives, and national development plans. They are not simply grassroots initiatives with individuals staking a claim to their national origins with recreational DNA tests. National biobanks are instruments that facilitate the basic scientific research of ethnic populations and foster the practice of performing national inclusion through the donation of biological samples.

The national biobanks in Israel and Qatar make claims to being "exceptional" by virtue of the unique genetic compositions of their populations. These claims, however, are not based on a high degree of homogeneous similarity, a unique purity, or a high degree of relatedness, but rather on the complex genetic diversity that characterizes both the Israeli and the Qatari populations. These claims demand recognition of the relationship between the utility of these claims and the society that has produced them. Utility, in the basic genetic research of populations, is somewhat proportionate to the diversity of the populations, particularly in the genomic study of inheritable disease.

Despite this general principle—that genetic diversity holds more scientific utility—the Israeli state must be understood in its historical context if one is to comprehend the role of ethnic genetics in this diverse society. From the early years of the Israeli state, Jewish nation-building faced several challenges, including the integration of Jewish immigrants from diverse backgrounds, the inclusion of indigenous Arab populations, and the balancing of power between secular and religious groups. These issues were tackled through strategies of immigration policy, like Israel's Law of Return, the development of a national language in modern Hebrew, and the establishment of a "civil religion," with Jewish national holidays and monuments. But more recently Israel has received many non-Jewish immigrants, raising controversy over who is a “real Jew.” Divisions between Orthodox and secular Jews have also increased, and Israel has moved away from its socialist beginnings to become a more unequal, market-driven society so that the divide between the political right and left now primarily distinguishes supporters of territorial expansion from those who favor return to the pre-1967 territories and support in principle the establishment of a Palestinian state. Within this fraught context, the “‘biologization’ of Jewish culture and historical narrative” (Egorova 2014, 354) affords the possibility of imagining continuity, solidarity, and collectivity, when sociological divides could otherwise be emphasized to the detriment of national cohesiveness.

A comparable mode of unification is at play in Qatar, where tribal familial identity is giving way to an emergent national identity, at least at the formal level of state symbols and public culture, most effectively exemplified by National Day, the Msheireb museum complex, the newly opened National Museum, QB, and the Qatar Genome Programme. This is not to say that old alliances and family “bloodlines” have eroded (Fromherz 2012, 113). Rather, Qatar now enjoys peaceful neo-traditionalism, coupling tribal traditions and modern consumer-driven lifestyles. This historical movement has carried with it an imagination of national collectivity that is taking powerful symbolic form in national biomedical developments.

Both the Qatari and Israeli instantiations of ethnic genetics and biological peoplehood demand recognition of the relationships between basic science and the national context, with a specific focus on the role of state institutions in driving these efforts. The phenomenon can be thought of broadly as the molecularization of identity, which I gloss as *the privileging of the molecular realm as a site for authoritative articulations of ethnonational identity and belonging*. The broader national context thus becomes *reified* content in the biosciences. The national context is reified in the sense that the nation as metaphysical imaginary takes concrete form in the misrecognized relations between the citizenry and its supplement, the excluded, in the imagination of genomic citizenship.

This process is significant for the anthropology of science, the anthropology of identity and belonging, and for science, technology, and society, as it addresses ethnic and national identity as they are reaffirmed through biological technologies. While this phenomenon of genomic citizenship I am describing here is influenced by many factors, fundamentally it is the entanglement of the technical, the political, and the epistemic that makes it possible to apprehend the nation as a natural biological entity. The national collective can be apprehended, treated, and managed as a biological object. The national imaginary begets—is coproduced with—a national object in the form of the bio-nation. Perhaps this marks a new moment in the history of Middle Eastern nationalisms, and of nationalisms in general.

In juxtaposing the particular biopolitics of Qatar and Israel, however, this book has brought to light the distinct ways that global scientific trends may be localized and incorporated into varying ethnonational imaginaries and the entailed political projects of inclusion and exclusion. In both states, we see the genome mediating the imagination of citizenship and fostering hopes of healthy and prosperous futures.

This phenomenon of genomic citizenship is part of building states that are networked into the global economy of knowledge production toward the imagined goal of a new era of precision medicine. Global science thus becomes localized. Such universal mobile science thereby reinforces the mythical ethnic identities that are the precondition of the ethnonation that makes this movement itself possible. Science, curiously, both dispels and reiterates the mythological.

NOTES

CHAPTER 1

1. According to the National Transplant Center, the center "was established by the authority of the Ministry of Health in 1994, with the purpose of creating an official and independent body for the management and coordination of organ donation and transplantation in Israel" (2018).

2. See Israel Organ Transplant Act (2008); Mor and Boas (2005); National Transplant Center (2018). Signing an Adi organ donor card expresses the willingness of the holder to donate their organs after death, to help save the lives of patients waiting for an organ transplant. The names of signatories to the Adi card system are deposited in a confidential database, and possession of a card grants priority to the holder on the transplant waiting list, and also to their close relatives, should they need a transplant. Both these established donor systems already merge neoliberal market logics (foregrounding individual choice) with altruistic values and the participatory ethics and solidarity of a collectivist society. In other words, participants gain the option of personally benefiting from their contribution, but it nonetheless remains more likely that individual contributions will help others. In terms of how they work, these systems may be similar to the emergent personalized medical models, in which individuals could volunteer personal data in order to be accorded both direct benefits, by way of access to personal health assessment, and indirect benefits, by helping the wider community become healthier.

CHAPTER 2

1. See Prainsack and Hashiloni-Dolev (2009) for a review of the impacts of the so-called new genetics on collective identities, including nation, race, and ethnicity.

2. See Alperin (2014); Wheelwright (2013).

3. The Jewish holy temple that stood on the Temple Mount in Jerusalem and that the Romans destroyed in 70 CE.

4. See Nesis (1970, 59), quoting Eban (1984, 191): "The driving force in Israel's life is still generated by immigration movements."

5. "The hebrew word *leoum* can be translated as 'ethnic group' or 'nationality' or 'peoplehood'" (Nesis 1970, 54).

6. Israel's Population Registry Law, 5725-1965, 19 LSI 288 (1964–1965) (Isr.) (replacing the Registration of Inhabitants Ordinance, 5709-1949).

7. See Gross (2013); CA 8573/08 Ornan et al. v. Ministry of Interior (Oct. 2, 2013, amended on June 10, 2013), (Isr.); HCJ 8140/13 Ornan v. State of Israel (Dec. 9, 2013) (Isr.) ("We are dealing here with a sensitive and highly controversial issue on both a historical and moral level that has been with the Jewish people for many years and with the Zionist movement from its very beginnings. The concept that Judaism is not merely a religion but also a national affiliation is a cornerstone of Zionism. Against it presents itself the concept according to which Judaism is merely a religion, and therefore the national affiliation of Jews is according to the state of which they are citizens.") See also CA 630/70 Tamarin v. State of Israel, 26(1) PD 197 [1972] (Isr.) (rejecting Tamarin's subjective feeling of belonging in the Israeli nation and refusing to let Tamarin "change the entry of the rubric *le'om* in his identity card and in his file in the Registry from Jewish to Israeli").

8. Basic Law: Israel—The Nation State of the Jewish People, https://knesset.gov.il/laws/special/eng/BasicLawNationState.pdf.

9. Law of Return, 5710-1950, 4 LSI 114 (1949–1950) (Isr.).

10. Law of Return (Amendment 5714-1954), SH No. 163 p. 174 (Isr.).

11. Law of Return (Amendment 5730-1970), SH No. 586 p. 34 (Isr.).

12. Law of Return (Amendment 5730-1970), SH No. 586 p. 34 (Isr.).

13. It is important to note that the amended law did not define what type of conversion was necessary. From its enactment, it was identified as an area for future challenge: "from [that] point onwards, the question was no longer 'who is a Jew'; it became instead 'who is a convert'" (Sapir 2006, 39).

14. The term "seed of Israel" "also has a [slightly different and] broader definition that applies to anyone with demonstrated Jewish ancestry dating back several generations" (Maltz 2015).

15. Shalit v. Minister of the Interior involved a Jewish naval officer who married a non-Jewish Scottish woman; the couple lived in Israel with their two children (Baer 1971, 133). The Shalits were atheists and attempted to register their children as Jewish under the *le'oum* or nationality designation and leave the religion category blank (133–134). The registry, under guidance from the Ministry of Interior, refused to permit this, since Mrs. Shalit was not Jewish, and therefore the children did not belong to the Jewish nation under religious law (134). The Israeli Supreme Court initially attempted to avoid a decision and recommended that the Knesset strike the *le'oum* or ethnic category from the Registry Law because it was too vague (134). However, the Knesset disregarded the request, and the Court was forced to decide (134). In an unprecedented nine-judge panel decision, delivered via eight long opinions, the court ruled 5 to 4 that the ministry clerk did not have the right to

question the Shalits' application and thus the children should be registered as their parents wished (135). The majority attempted to limit the scope of the decision; Justice Sussman in concurrence explained that "the question is not 'Who is a Jew?,' since the term has many meanings, but rather who is considered a Jew for purposes of this law" (142). Justice Cohen wrote that this was a secular law that the court was asked to interpret and therefore religious law should not control (142). Further, he qualified the decision by reiterating that the Registry Act "states that the answers to *leum* and religion do not provide prima facie evidence of their correctness" (142). Almost immediately, in response, the Knesset amended the Law of Return and the Population Registry Law to mandate that anyone who registers as Jewish under either nationality or religion classifications must meet the religious definition (145). As one commentator described it, "The amended law 'overruled' the Shalit case by adopting the religious law test of defining who is considered Jewish, but the law saved the spirit of the Shalit decision" by granting non-Jewish family members the right to immigrate under the Law of Return (Altschul 2002, 1357).

16. Burton (2015, 79), citing Cohen and Susser (2009).

17. *The Law of Return*, The Jewish Agency, http://www.jewishagency.org/first-steps/program/5131 (accessed February 16, 2015).

18. See, e.g., Richmond (1993).

19. Hammer (2011). One reporter noted that several hundreds of thousands of immigrants came from the FSU in the 1990s alone on the basis that they were "seeds of Israel." See also Altschul (2002, 1359) (estimating that more than half of the Russian immigrants who arrived in 1999 were not Jewish under religious law).

20. Richmond (1993, 117): "estimating that in the 1990s three percent of Russian immigrants abuse the system in this manner."

21. See Hammer (2011), 1.

22. See Hammer (2011), 11.

23. This is according to Rabbinical Court Jurisdiction (Marriage and Divorce) Law, 1953 §1, 7 LSI 139 (1953) (Isr.). Although adoption and inheritance used to be under the religious court jurisdiction, this was changed through various legislative acts. It was the result of the Law and Administration Ordinance, 1948, 1 LSI 9 (1948) (Isr.).

24. The Declaration of the Establishment of the State of Israel, May 14, 1948.

25. See, e.g., Rabbi Dr. Lawrence S. Nesis (1970, 53): "The question of who is a Jew had long been the subject of controversy in Israel"; and Mark J. Altschul (2002, 1352): "Defining who is Jewish by Israeli standards is perhaps the most difficult question that has faced Israel since its inception."

26. See Richmond (1993, 100), for an analysis of Israel's Law of Return.

27. See Falk (2017) for a comprehensive overview of the entangled history of Zionism and the biology of Jews.

28. The Beta Israel are Ethiopian Jews who mostly immigrated to Israel in the 1980s and 1990s (see Seeman 2010). The Kuki-Chin-Mizo is a small group that claims to be descended from the tribe of Menashe.

29. See Abu El-Haj (2012, 287); Kahn (2010, 13); Skorecki et al. (1997); Thomas et al. (2000, 2002).

30. See Abu El-Haj (2012) for a comprehensive review of the science of Jewish origins.

CHAPTER 3

1. See Troen and Troen (2019) for a discussion of the concept of indigeneity in debates over belonging in Israel/Palestine.

2. "In 2017, EuroBioBank Network is composed of 25 rare disease biobank members from 9 European countries (France, Germany, Hungary, Italy, Malta, Slovenia, Spain, United Kingdom and Turkey) as well as Israel and Canada" (Euro BioBank 2020).

3. For example, each growing cell line costs US$150, the price for a 10 μg DNA sample is US$45, and a 5 μg DNA sample costs US$30 (excluding shipping charges) (NLGIP 2019b).

4. I recognize that the geographical limits of the state of Israel are difficult to determine. For the purposes of this conversation, I am assuming that the population of the state of Israel is the population within the internationally recognized "green line," not including Israeli settlements and residents in the disputed territories.

5. I use "ethnic genes" to denote the diverse practices, measurements, and claims about populations and the natural associations of individuals. I do not wish to fall into the trap of equating the signifier "ethnic genes" with the genetic sequences that are putatively ethnic markers. "Ethnic genes" therefore means both the process of reification of genes and elective human identities and their mutual conflation in the practices of population genetics. "Ethnic genetics" stands for the ontological mediations that render misrecognition possible.

CHAPTER 4

1. The Functional Genomics Symposium's aim was to discuss advances in functional genomics and genomic medicine, to address the impact of genetic studies on complex disorders and rare diseases, and to share new knowledge of functional mapping of the human genome as it relates to "precision medicine." Technology and service providers made presentations in parallel with the symposium presentations. The symposium was organized as part of the Sidra Symposium Series and was aimed at academics, researchers, physicians, health-care providers, and regulatory agents working in the fields of genomics and genomic medicine.

2. As the literature on the history and culture of Qatar is very sparse, I draw primarily on the work of Fromherz (2012), which offers a comprehensive overview of the historical emergence of Qatari national identity.

3. Breastfeeding, for example, is a key issue for Qatar, and during World Breastfeeding Week 2015, Sidra issued an infographic to help women be able to both breastfeed and return to

work. See http://www.sidra.org/wp-content/uploads/2015/08/sidra-breastfeeding-infographic-english.pdf.

4. Hamad Medical Corporation (HMC) "has been the principal public healthcare provider in the State of Qatar for over three decades, and is dedicated to delivering the safest, most effective and most compassionate care to each and every one of our patients. HMC manages eight hospitals, incorporating five specialist hospitals and three community hospitals. . . . While HMC continues to upgrade its facilities and services, it has also embarked on an ambitious expansion program, targeting the areas of need in our community." Hamad Medical Corporation (2015).

CHAPTER 5

1. Personal interview, Doha, Qatar, December 12, 2018.
2. Genomics England was set up in 2015 to deliver the 100,000 Genomes Project, its flagship project to sequence 100,000 whole genomes from National Health Service patients with rare diseases, and their families, as well as patients with common cancers. In late 2018, the secretary of state for health and social care, the Rt. Hon. Matt Hancock MP, announced plans to sequence five million genomes over the following five years (Genomics England 2018).
3. Much of the scholarship on this topic has focused on the ethics of sampling and storage of human biological material and medical information (Cambon-Thomsen 2004; Cambon-Thomsen et al. 2007; Haga and Beskow 2008; Hansson 2009; McGonigle and Shomron 2016), and other work has looked at the social problems and limitations of collective and individual consent (Caulfield and Kaye 2009; Gottweis, Gaskell, and Starkbaum 2011; Hansson et al. 2006) and the entailed protection of personal data and the legal definition of the nature of the individual participant (Gurwitz 2015). There is also a growing awareness of the issues at stake in transnational collaborations that entail challenges for governance where different regulatory and ethical regimes make cross-border harmonization difficult (Chen 2013; Gottweis and Lauss 2012; Gottweis and Petersen 2008; Kaye 2011).
4. Personal interview, Doha, Qatar, December 11, 2018.
5. An Islamic legal opinion, given by a qualified jurist.

CHAPTER 6

1. This can be said more forcefully since July 2018 when the Israeli parliament, the Knesset, passed a basic law to define Israel as the nation-state of the Jewish people. See "Basic Law: Israel—The Nation State of the Jewish People," https://knesset.gov.il/laws/special/eng/BasicLawNationState.pdf.

REFERENCES

Abu El-Haj, Nadia. 2001. *Facts on the Ground: Archaeological Practice and Territorial Self-Fashioning in Israeli Society.* Chicago: University of Chicago Press.

Abu El-Haj, Nadia. 2007a. The Genetic Reinscription of Race. *Annual Review of Anthropology* 36, 283–300.

Abu El-Haj, Nadia. 2007b. Rethinking Genetic Genealogy: A Response to Stephan Palmié. *American Ethnologist* 34(2), 223–226.

Abu El-Haj, Nadia. 2012. *The Genealogical Science: The Search for Jewish Origins and the Politics of Epistemology*. Chicago: University of Chicago Press.

Adorno, Theodor. 1980 [1966]. *Negative Dialectics*. 2nd ed. London: Bloomsbury.

Ahad Ha'Am. 1898. The Transvaluation of Values. In *Selected Essays by Ahad Ha'Am*, edited and translated by Leon Simon, 217–241. Philadelphia: The Jewish Publication Society of America.

Ahad Ha'Am. 1904. Flesh and Spirit. In *Selected Essays by Ahad Ha'Am*, edited and translated by Leon Simon, 139–216. Philadelphia: The Jewish Publication Society of America.

Almog, Oz. 2000. *The Sabra: The Creation of the New Jew*. Translated by Haim Watzman. Los Angeles: University of California Press.

Alperin, Michele. 2014. How DNA Testing Can Reveal Jewish Ancestry, Bolster Zionist Narrative. Jewish News Syndicate, October 15. http://www.jns.org/latest-articles/2014/9/19/how-dna-testing-can-reveal-jewish-ancestry-and-bolster-the-zionist-narrative.

Altschul, Mark J. 2002. Israel's Law of Return and the Debate of Altering, Repealing, or Maintaining Its Present Language. *University of Illinois Law Review* 2002(5), 1345–1372.

Anderson, Benedict. 1983. *Imagined Communities*. London: Verso.

Anderson, Perry. 2015. The House of Zion. *New Left Review* 96(11/12), 5–37.

Appadurai, Arjun. 1990. Disjuncture and Difference in the Global Cultural Economy. *Theory Culture Society* 7(2), 295–310.

Atzmon, Gil, Li Hao, Itsik Pe'er, Christopher Velez, Alexander Pearlman, Pier Francesco Palamara, Bernice Morrow, et al. 2010. Abraham's Children in the Genome Era: Major Jewish Diaspora Populations Comprise Distinct Genetic Clusters with Shared Middle Eastern Ancestry. *American Journal of Human Genetics* 86(6), 850–859.

Baer, Noah. 1971. Casenotes: Who Is a Jew? A Determination of Ethnic Status for Purposes of the Israeli Population Registry Act. *Columbia Journal of Transnational Law* 10, 133–145.

Barak, Oren. 2002. Intra-communal and Inter-communal Dimensions of Conflict and Peace in Lebanon. *International Journal of Middle Eastern Studies* 34(4), 619–644.

Barber, Benjamin R. 1992. Jihad vs. McWorld. *The Atlantic* 269(3), 53–65.

Barell, Ari, and David Ohana. 2014. "The Million Plan": Zionism, Political Theology and Scientific Utopianism. *Politics, Religion & Ideology* 15(1), 1–12.

BBC *HARDtalk*. 2019. Gebran Bassil, September 21.

Behar, Doron M., Michael F. Hammer, Daniel Garrigan, Richard Villems, Batsheva Bonne-Tamir, Martin Richards, David Gurwitz, et al. 2004. mtDNA Evidence for a Genetic Bottleneck in the Early History of the Ashkenazi Jewish Population. *European Journal of Human Genetics* 12(5), 355–364.

Behar, Doron M., Ene Metspalu, Toomas Kivisild, Alessandro Achilli, Yarin Hadid, Shay Tzur, Luisa Pereira, et al. 2006. The Matrilineal Ancestry of Ashkenazi Jewry: Portrait of a Recent Founder Event. *American Journal of Human Genetics* 78(3), 487–497.

Behar, Doron M., Ene Metspalu, Toomas Kivisild, Saharon Rosset, Shay Tzur, Yarin Hadid, Guennady Yudkovsky, et al. 2008. Counting the Founders: The Matrilineal Genetic Ancestry of the Jewish Diaspora. *PLOS One* 3(4), e2062.

Behar, Doron M., Bayazit Yunusbayev, Mait Metspalu, Ene Metspalu, Saharon Rosset, Jüri Parik, Siiri Rootsi, et al. 2010. The Genome-Wide Structure of the Jewish People. *Nature* 466(7303), 238–242.

Benjamin, Ruha. 2009. A Lab of Their Own: Genomic Sovereignty as Postcolonial Science Policy. *Policy & Society* 28(4), 341–355.

Berdichevski, Micha Josef. 1997. Wrecking and Building (1900). In *The Zionist Idea*, edited by Arthur Hertzberg, 293–295. Philadelphia: The Jewish Publication Society.

Berman, Lila Corwin. 2009. *Speaking of Jews: Rabbis, Intellectuals, and the Creation of an American Public Identity*. Los Angeles: University of California Press.

Bijker, Wiebe, Thomas P. Hughes, and Trevor Pinch, eds. 1987. *The Social Construction of Technological Systems: New Directions in the Sociology and History of Technology*. Cambridge, MA: MIT Press.

Birenbaum-Carmeli, Daphna. 2009. The Politics of "The Natural Family" in Israel: State Policy and Kinship Ideologies. *Social Science and Medicine* 69(7), 1018–1024. doi:10.1016/j.socscimed.2009.07.044.

Birthright Israel. 2015. Taglit-Birthright Israel, FAQ. Accessed April 30, 2015. http://www.birthrightisrael.com/Pages/Help-Center-Answers.aspx?ItemID=1.

Bloom, Etan. 2007. What "The Father" Had in Mind? Arthur Ruppin (1876–1943), Cultural Identity, Weltanschauung and Action. *History of European Ideas* 33(3), 330–349.

Bloor, David. 1991. *Knowledge and Social Imagery*. Chicago: University of Chicago Press.

Bourdieu, Pierre. 2001. *Science of Science and Reflexivity*. Cambridge: Polity Press.

Boyarin, Daniel, and Jonathan Boyarin. 1993. Generation and the Ground of Jewish Identity. *Critical Inquiry* 19(4), 693–725.

Braverman, Irus. 2014. *Planted Flags: Trees, Land, and Law in Israel/Palestine*. Cambridge: Cambridge University Press.

Bray, Steven M., Jennifer G. Mulle, Anne F. Dodd, Ann E. Pulver, Stephen Wooding, and Stephen T. Warren. 2010. Signatures of Founder Effects, Admixture, and Selection in the Ashkenazi Jewish Population. *Proceedings of the National Academy of Sciences USA* 107(37), 16222–16227.

Burton, Elise K. 2015. An Assimilating Majority? Israeli Marriage Law and Identity in the Jewish State. *Journal of Jewish Identities* 8(1), 73–94.

Burton, Elise K. 2018. Narrating Ethnicity and Diversity in Middle Eastern National Genome Projects. *Social Studies of Science* 48(5), 762–786.

Busby, Helen, and Paul Martin. 2006. Biobanks, National Identity and Imagined Communities: The Case of UK Biobank. *Science as Culture* 15(3), 237–251.

Business in Qatar and Beyond. 2013. Population of Qatar by Nationality. *BQ Magazine*. Accessed November 22, 2015. http://www.bq-magazine.com/economy/2013/12/population-qatar.

Cambon-Thomsen, Anne. 2004. The Social and Ethical Issues of Post-Genomic Human Biobanks. *Nature Reviews Genetics* 5(11), 866–873.

Cambon-Thomsen, Anne, Pascal Ducournau, Pierre-Antoine Gourraud, and David Pontille. 2003. Biobanks for Genomics and Genomics for Biobanks. *Comparative and Functional Genomics* 4(6), 628–634. doi:10.1002/cfg.333.

Cambon-Thomsen, Anne, Emmanuelle Rial-Sebbag, and Bartha M. Knoppers. 2007. Trends in Ethical and Legal Frameworks for the Use of Human Biobanks. *European Respiratory Journal* 30(2), 373–382.

Caulfield, Timothy, and Jane Kaye. 2009. Broad Consent in Biobanking: Reflections on Seemingly Insurmountable Dilemmas. *Medical Law International* 10(2), 85–100.

Central Bureau of Statistics. 2019. Population of Israel on the Eve of 2020. Cbs.Gov.Il. https://www.cbs.gov.il/he/mediarelease/DocLib/2019/413/11_19_413e.pdf.

Chadwick, Ruth, and Kåre Berg. 2001. Solidarity and Equity: New Ethical Frameworks for Genetic Databases. *Nature Reviews Genetics* 2(4), 318–321.

Chen, Haidan. 2013. Governing International Biobank Collaboration: A Case Study of China Kadoorie Biobank. *Science, Technology & Society* 18(3), 321–338.

Chernick, Ilanit. 2017. Should Jewishness Be Determined by a Genetic Test? *Jerusalem Post*, November 25. https://www.jpost.com/Magazine/Should-Jewishness-be-determined-by-a-genetic-test-514968.

Chesler, Caren. 2013. What Makes a Jewish Mother. *New York Times*, June 3. https://well.blogs.nytimes.com/2013/06/03/what-makes-a-jewish-mother.

Chu, Julie Y. 2010. *Cosmologies of Credit: Transnational Mobility and the Politics of Destination in China*. Durham, NC: Duke University Press.

CIA World Factbook. 2018. Rank Order. In *CIA World Factbook*. Accessed February 27, 2018. https://www.cia.gov/library/publications/the-world-factbook/rankorder/2004rank.html#qa.

CIA World Factbook. 2020a. Israel. In *CIA World Factbook*. Accessed July 23, 2020. https://www.cia.gov/library/publications/the-world-factbook/geos/is.html.

CIA World Factbook. 2020b. Qatar. In *CIA World Factbook*. Accessed February 12, 2020. https://www.cia.gov/library/publications/the-world-factbook/geos/qa.html.

Cohen, Asher, and Bernard Susser. 2009. Jews and Others: Non-Jewish Jews in Israel. *Israel Affairs* 15(1), 52–65.

Comaroff, Jean, and John Comaroff. 2009. *Ethnicity Inc.* Chicago: University of Chicago Press.

Comaroff, Jean, and John Comaroff. 2016. *The Truth about Crime: Sovereignty, Knowledge, Social Order*. Chicago: University of Chicago Press.

Comaroff, John, and Jean Comaroff. 1992. *Ethnography and the Historical Imagination*. Boulder, CO: Westview Press.

Cook, Michael. 2016. Kuwait Becomes First Country to Demand Universal DNA Tests. *BioEdge*, August 27. https://www.bioedge.org/bioethics/kuwait-becomes-first-country-to-demand-universal-dna-tests/11974#disqus_thread.

Cooper, Melinda. 2008. *Life as Surplus: Biotechnology and Capitalism in the Neoliberal Era*. Seattle: University of Washington Press.

Daston, Lorraine. 2000. *Biographies of Scientific Objects*. Edited by Lorraine Daston. Chicago: University of Chicago Press.

Descola, Philippe. 2013. *Beyond Nature and Culture*. Chicago: University of Chicago Press.

Dominguez, Virginia R. 1989. *People as Subject, People as Object: Selfhood and Peoplehood in Contemporary Israel*. Madison: University of Wisconsin Press.

Dor Yeshorim. 2015. Ashkenazi Genetic Traits. Accessed April 29, 2015. https://www.jewishgenetics.org/ashkenazi-genetic-traits.

Dumit, Joseph. 2012. *Drugs for Life*. Durham, NC: Duke University Press.

Durkheim, Émile. 2008 [1912]. *The Elementary Forms of Religious Life*. Edited by Mark S. Cladis and translated by Carol Cosman. Oxford: Oxford University Press.

Eban, Abba. 1984. *My People: The Story of the Jews*. New York: Random House.

Efron, John M. 1994. *Defenders of the Race: Jewish Doctors and Race Science in Fin-de-Siècle Europe.* New Haven, CT: Yale University Press.

Egorova, Yulia. 2014. Theorizing "Jewish Genetics": DNA, Culture, and Historical Narrative. In *The Routledge Handbook of Contemporary Jewish Cultures*, edited by Nadia Valman and Laurence Roth, 353–364. London: Routledge.

Egorova, Yulia, and Shahid Perwez. 2010. The Children of Ephraim: Being Jewish in Andhra Pradesh. *Anthropology Today* 26(6), 14–19.

Egorova, Yulia, and Shahid Perwez. 2012. Old Memories, New Histories: (Re)discovering the Past of Jewish Dalits. *History and Anthropology* 23(1), 1–15.

Eisenstein, Michael. 2015. Big Data: The Power of Petabytes. *Nature* 527(7576), S2–S4. doi:10.1038/527S2a.

Elhaik, Eran. 2012. The Missing Link of Jewish European Ancestry: Contrasting the Rhineland and the Khazarian Hypotheses. *Genome Biology and Evolution* 5(1), 61–74.

Epstein, Steven. 2007. *Inclusion: The Politics of Difference in Medical Research.* Chicago: University of Chicago Press.

Euro BioBank. 2020. About. Accessed November 19, 2020. http://www.eurobiobank.org/en/information/info_institut.htm.

European Science Foundation (ESF). 2012. *Personalised Medicine for the European Citizen—Towards More Precise Medicine for the Diagnosis, Treatment and Prevention of Disease.* Strasbourg: ESF.

Fakhro, Khalid A., Michelle R. Staudt, Monica Denise Ramstetter, Amal Robay, Joel A. Malek, Ramin Badii, Ajayeb Al-Nabet Al-Marri, et al. 2016. The Qatar Genome: A Population-Specific Tool for Precision Medicine in the Middle East. *Human Genome Variation* 3(16016).

Falk, Raphael. 1998. Zionism and the Biology of the Jews. *Science in Context* 11(34), 587–607.

Falk, Raphael. 2017. *Zionism and the Biology of Jews.* Cham, Switzerland: Springer International Publishing.

Feige, Michael. 2009. *Settling in the Hearts: Fundamentalism, Time, and Space in Judea and Samaria.* Detroit: Wayne State University Press.

Fortun, Michael. 2008. *Promising Genomics: Iceland and deCODE Genetics in a World of Speculation.* Berkeley: University of California Press.

Foucault, Michel. 1977. *Discipline and Punish: The Birth of the Prison.* New York: Pantheon Books.

Foucault, Michel. 2010. *The Birth of Biopolitics: Lectures at the Collège de France, 1978–1979.* London: Picador.

Franklin, Sarah. 1995. Science as Culture, Cultures of Science. *Annual Review of Anthropology* 24, 163–184.

Franklin, Sarah. 2007. *Dolly Mixtures: The Remaking of Genealogy.* Durham, NC: Duke University Press.

Fromherz, Allen James. 2012. *Qatar: A Modern History*. Washington, DC: Georgetown University Press.

Fujimura, Joan H., and Ramya Rajagopalan. 2011. Different Differences: The Use of "Genetic Ancestry" versus Race in Biomedical Human Genetic Research. *Social Studies of Science* 41(1), 5–30.

Fullwiley, Duana. 2008. The Biologistical Construction of Race: "Admixture" Technology and the New Genetic Medicine. *Social Studies of Science* 38(5), 695–735.

Gell, Alfred. 1998. *Art and Agency: An Anthropological Theory*. Oxford: Oxford University Press.

Genomics England. 2018. Genomics England. Accessed September 29, 2019. https://www.genomicsengland.co.uk/.

Ghaly, Mohammed, Eman Sadoun, Fowzan Alkuraya, Khalid Fakhro, Ma'n Zawati, Said Ismail, and Tawfeg Ben-Omran. 2016. *Genomics in the Gulf Region and Islamic Ethics*. Edited by Mohammed Ghaly, World Innovation Summit for Health. Accessed May 19, 2019. https://www.wish.org.qa/wp-content/uploads/2018/01/Islamic-Ethics-Report-EnglishFINAL.pdf.

Goldstein, David B. 2009. *Jacobs Legacy: A Genetic View of Jewish History*. New Haven, CT: Yale University Press.

Goldstein, Eric L. 2006. *The Price of Whiteness: Jews, Race and American Identity*. Princeton, NJ: Princeton University Press.

Gordon, Aaron David. 1997a. People and Labor (1911). In *The Zionist Idea*, edited by Arthur Hertzberg, 372–374. Philadelphia: The Jewish Publication Society.

Gordon, Aaron David. 1997b. Our Tasks Ahead (1920). In *The Zionist Idea*, edited by Arthur Hertzberg, 381–382. Philadelphia: The Jewish Publication Society.

Gottweis, Herbert, George Gaskell, and Johanne Starkbaum. 2011. Connecting the Public with Biobank Research: Reciprocity Matters. *Nature Reviews Genetics* 12(11), 738–739.

Gottweis, Herbert, and Byoungsoo Kim. 2009. Bionationalism, Stem Cells, BSE, and Web 2.0 in South Korea: Toward the Reconfiguration of Biopolitics. *New Genetics and Society* 28(3), 223–239.

Gottweis, Herbert, and Georg Lauss. 2012. Biobank Governance: Heterogeneous Modes of Ordering and Democratization. *Journal of Community Genetics* 3(2), 61–72.

Gottweis, Herbert, and Alan Petersen. 2008. *Biobanks: Governance in Comparative Perspective*. Edited by Herbert Gottweis and Alan Petersen. London: Routledge.

Greene, Jeremy. 2014. *Generic: The Unbranding of Modern Medicine*. Baltimore: Johns Hopkins University Press.

Gross, Aeyal. 2013. Court Rejection of Israeli Nationality Highlights Flaws of Jewish Democracy. *Ha'Aretz*, October 3. http://www.haaretz.com/opinion/.premium-1.550336.

Gurwitz, David. 2015. Genetic Privacy: Trust Is Not Enough. *Science* 347(6225), 957–958.

Gurwitz, David, Orit Kimchi, and Batsheva Bonne-Tamir. 2003. The Israeli DNA and Cell Line Collection: A Human Diversity Repository. In *Populations and Genetics: Legal and Socio-Ethical Perspectives*, edited by Bartha Maria Knoopers, 95–113. Leiden: Martinus Nijhoff.

Hacking, Ian. 1995. The Looping Effects of Human Kinds. In *Causal Cognition: A Multidisciplinary Debate*, edited by Dan Sperber, David Premack, and Ann James Premack, 351–394. New York: Oxford University Press.

Hacking, Ian. 1998. *Rewriting the Soul.* Princeton, NJ: Princeton University Press.

Hacking, Ian. 2000. *The Social Construction of What?* Cambridge, MA: Harvard University Press.

Hacking, Ian. 2002. *Historical Ontology*. Cambridge, MA: Harvard University Press.

Hacking, Ian. 2006. Making Up People. *London Review of Books* 28(16), 23–26.

Haga, Susanne B., and Laura M. Beskow. 2008. Ethical, Legal, and Social Implications of Biobanks for Genetics Research. *Advances in Genetics* 60, 505–544.

Hamad Medical Corporation. 2015. About Us. Accessed November 28, 2015. http://www.qatarbiobank.org.qa/about-us/what-is-qatar-biobank.

Hammer, M. F., A. J. Redd, E. T. Wood, M. R. Bonner, H. Jarjanazi, T. Karafet, S. Santachiara-Benerecetti, et al. 2000. Jewish and Middle Eastern Non-Jewish Populations Share a Common Pool of Y-Chromosome Biallelic Haplotypes. *Proceedings of the National Academy of Sciences* 97(12), 6769–6774.

Hammer, Rabbi Reuven. 2011. On Proving Jewish Identity YD 268:10.2011. Committee on Jewish Law and Standards, CJLS, the Rabbinical Assembly, May 24, 1. http://www.rabbinicalassembly.org/jewish-law/committee-jewish-law-and-standards/yoreh-deah.

Hansson, M. G. 2009. Ethics and Biobanks. *British Journal of Cancer* 100(1), 8–12.

Hansson, Mats G., Joakim Dillner, Claus R. Bartram, Joyce A. Carlson, and Gert Helgesson. 2006. Should Donors Be Allowed to Give Broad Consent to Future Biobank Research? *Lancet Oncology* 7(3), 266–269.

Hart, Mitchell B. 1999. Racial Science, Social Science, and the Politics of Jewish Assimilation. *Isis* 90(2), 268–297.

Hart, Mitchell B. 2000. *Social Science and the Politics of Modern Jewish Identity*. Palo Alto, CA: Stanford University Press.

Hart, Mitchell B. 2011. *Jews and Race: Writings on Identity and Difference*. Waltham, MA: Brandeis University Press.

Hayden, Cori. 2003. *When Nature Goes Public: The Making and Unmaking of Bioprospecting in Mexico*. Princeton, NJ: Princeton University Press.

Heathcote, Edwin. 2019. The New National Museum of Qatar Is a Desert Rose of Mutant Scale. *Financial Times*, March 28. https://www.ft.com/content/622127a6-507d-11e9-b401-8d9ef1626294.

Helmreich, Stefan. 2007. An Anthropologist Underwater: Immersive Soundscapes, Submarine Cyborgs, and Transductive Ethnography. *American Ethnologist* 34(4), 621–641.

Herzl, Theodor. 1896. *The Jewish State*. Accessed May 19, 2019. http://www.MidEastweb.org.

Hess, Jonathan M. 2002. *Germans, Jews and the Claims of Modernity*. New Haven, CT: Yale University Press.

Hinterberger, Amy. 2012. Publics and Populations: The Politics of Ancestry and Exchange in Genome Science. *Science as Culture* 21(4), 528–549.

Hirsch, Dafna. 2009. Zionist Eugenics, Mixed Marriage, and the Creation of a "New" Jewish Type. *Journal of the Royal Anthropological Institute* 15(3), 592–609.

Hogle, Linda F. 1999. *Recovering the Nation's Body: Cultural Memory, Medicine, and the Politics of Redemption*. New Brunswick, NJ: Rutgers University Press.

Horkheimer, Max, and Theodore Adorno. 2002 [1947]. *Dialectic of Enlightenment*. Palo Alto, CA: Stanford University Press.

Human Longevity Inc. (HLI). 2018. Your Health Intelligence Partner. Accessed March 6, 2018. http://www.humanlongevity.com.

Inhorn, Marcia C. 2015. *Cosmopolitan Conceptions: IVF Sojourns in Global Dubai*. Durham, NC: Duke University Press.

InnVentis. 2015. Home. Accessed November 19, 2015. http://www.innventis-pharma.com.

Israel Organ Transplant Act. 2008. Israel Transplant Law Organ Transplant Act. *Declaration of Istanbul.* Accessed November 4, 2015. http://www.declarationofistanbul.org/resources/legislation/267-israel-transplant-law-organ-transplant-act-2008#.

Jabloner, Anna. 2015. Humanity Pending: Californian Genomics and the Politics of Biology. PhD diss., Department of Anthropology, University of Chicago.

Jasanoff, Sheila. 2004. *States of Knowledge: The Co-production of Science and the Social Order*. London: Routledge.

Jasanoff, Sheila. 2005. *Designs on Nature: Science and Democracy in Europe and the United States*. Princeton, NJ: Princeton University Press.

Jasanoff, Sheila. 2011. *Reframing Rights: Bioconstitutionalism in the Genetic Age*. Cambridge, MA: MIT Press.

Jasanoff, Sheila, and Sang-Hyun Kim. 2013. Sociotechnical Imaginaries and National Energy Policies. *Science as Culture* 22(2), 189–196.

Jasanoff, Sheila, and Sang-Hyun Kim. 2015. Future Imperfect: Science, Technology, and the Imaginations of Modernity. In *Dreamscapes of Modernity: Sociotechnical Imaginaries and the Fabrication of Power*, edited by Sheila Jasanoff and Sang-Hyun Kim, 1–33. Chicago: University of Chicago Press.

Jasanoff, Sheila, Gerald E. Markle, James C. Peterson, and Trevor J. Pinch. 1994. *Handbook of Science and Technology Studies*. Edited by Sheila Jasanoff, Gerald E. Markle, James C. Peterson, and Trevor J. Pinch. Thousand Oaks, CA: Sage.

Jones, Justin. 2013. Damien Hirst Unveils Provocative Birth Sculptures in Doha. *The Daily Beast*, October 11. http://www.thedailybeast.com/articles/2013/10/11/damien-hirst-s-controversial-sculptures-in-doha.html.

Kahn, Susan Martha. 2005. The Multiple Meanings of Jewish Genes. *Culture, Medicine and Psychiatry* 29(2), 179–192.

Kahn, Susan Martha. 2010. Are Genes Jewish: Conceptual Ambiguities in the New Genetic Age. In *The Boundaries of Jewish Identity*, edited by Susan A. Glenn and Naomi B. Sokoloff, 12–26. Seattle: University of Washington Press.

Kaiser, Jocelyn. 2016. When DNA and Culture Clash: Saudi Arabia Is Making a Big Push into Human Genomics, Hoping to Prevent Inherited Diseases. *Science* 354(6317), 1217–1221.

Katriel, Tamar. 2004. *Dialogic Moments: From Soul Talks to Talk Radio in Israeli Culture*. Detroit: Wayne State University Press.

Kaye, Jane. 2004. Abandoning Informed Consent: The Case of Genetic Research in Population Collections. In *Genetic Databases: Socio-Ethical Issues in the Collection and Use of DNA*, edited by Oonagh Corrigan and Richard Tutton, 117–138. London: Routledge.

Kaye, Jane. 2011. From Single Biobanks to International Networks: Developing E-Governance. *Human Genetics* 130(3), 377–382.

Kaye/Kantrowitz, Melanie. 2007. *The Colors of Jews: Racial Politics and Racial Diasporism*. Bloomington: Indiana University Press.

Kimmerling, Baruch. 2005. *The Invention and Decline of Israeliness: State, Society, and the Military*. Oakland: University of California Press.

Kirsh, Nurit. 2003. Genetics in Israel in the 1950s: The Unconscious Internalization of Ideology. *Isis* 94(4), 631–655.

Knoopers, Bartha Maria. 2003. *Populations and Genetics Legal and Socio-Ethical Perspectives*. Leiden: Martinus Nijhoff.

Koestler, Arthur. 1976. *The Thirteenth Tribe: The Khazar Empire and Its Heritage*. New York: Random House.

Kohler, Noa Sophie. 2014. Genes as a Historical Archive: On the Applicability of Genetic Research to Sociohistorical Questions; The Debate on the Origins of Ashkenazi Jewry Revisited. *Perspectives in Biology and Medicine* 57(1), 105–117.

Kolata, Gina, and Heather Murphy. 2018. The Golden State Killer Is Tracked through a Thicket of DNA, and Experts Shudder. *New York Times*, April 27. https://www.nytimes.com/2018/04/27/health/dna-privacy-golden-state-killer-genealogy.html.

Kook, Rabbi Abraham Isaac. 1997. The Land of Israel (1910). In *The Zionist Idea*, edited by Arthur Hertzberg, 419–421. Philadelphia: The Jewish Publication Society.

Kravel-Tovi, Michal. 2015. Corrective Conversion: Unsettling Citizens and the Politics of Inclusion in Israel. *Journal of the Royal Anthropological Institute* 21(1), 127–146.

Latour, Bruno. 1987. *Science in Action: How to Follow Scientists and Engineers through Society*. Cambridge, MA: Harvard University Press.

Latour, Bruno. 1993. *We Have Never Been Modern*. Cambridge, MA: Harvard University Press.

Latour, Bruno. 2004. *Politics of Nature: How to Bring the Sciences into Democracy*. Cambridge, MA: Harvard University Press.

Latour, Bruno. 2005. *Reassembling the Social: An Introduction to Actor-Network Theory*. Oxford: Oxford University Press.

Latour, Bruno, and Steve Woolgar. 1986. *Laboratory Life: The Construction of Scientific Facts*. Princeton, NJ: Princeton University Press.

Lazar, Sian. 2013. *The Anthropology of Citizenship: A Reader*. London: Wiley-Blackwell.

Lebovic, Nitzan. 2015. Biometrics, or the Power of the Radical Center. *Critical Inquiry* 41(4), 841–868.

Lévi-Strauss, Claude. 1971. *Totemism*. Translated by Rodney Needham. Boston: Beacon Press.

Lewis, R. 2015. Precision Medicine: Much More Than Just Genetics. *PLOS Blogs*. Accessed May 19, 2019. http://blogs.plos.org/dnascience/2015/09/24/precision-medicine-medical-genetics/.

Magen David Adom in Israel. 2015. About Us. Accessed September 20, 2015. https://www.mdais.org.

Maltz, Judy. 2014. Jewish Enough for Birthright—But Not for Israel. *The Forward*, January 23. https://forward.com/news/israel/191456/jewish-enough-for-birthright-but-not-for-israel.

Maltz, Judy. 2015. How a Former Netanyahu Aide Is Boosting Israel's Jewish Majority, One "Lost Tribe" at a Time. *Ha'Aretz*, February 19. https://www.haaretz.com/how-one-man-is-pumping-up-israel-s-jewish-majority-1.5308899.

Marks, Jonathan. 2013. The Nature/Culture of Genetic Facts. *Annual Review of Anthropology* 42, 247–267.

Marx, Vivien. 2015. The DNA of a Nation. *Nature* 524(7566), 503–505.

Mathew, Lisa S., Manuel Spannagl, Ameena Al-Malki, Binu George, Maria F. Torres, Eman K. Al-Dous, Eman K. Al-Azwani, et al. 2014. A First Genetic Map of Date Palm (*Phoenix dactylifera*) Reveals Long-Range Genome Structure Conservation in the Palms. *BMC Genomics* 15(1), 285.

Mathew, Sweety, Susanne Krug, Thomas Skurk, Anna Halama, Antonia Stank, Anna Artati, Cornelia Prehn, et al. 2014. Metabolomics of Ramadan Fasting: An Opportunity for the Controlled Study of Physiological Responses to Food Intake. *Journal of Translational Medicine* 12(1), 161.

McGonigle, Ian. 2020a. Biobanking and "Qatarization": Ethno-national Identity in the Molecular Realm. In *Studies on the Social Construction of Identity and Authenticity*, edited by J. Patrick Williams and Kaylan C. Schwarz, 156–170. London: Routledge.

McGonigle, Ian. 2020b. National Biobanking in Qatar and Israel: Tracing How Global Scientific Institutions Mediate Local Ethnic Identities. *Science, Technology and Society*. doi:10.1177/0971721820931995.

McGonigle, Ian, and Stephan C. Schuster. 2019. Global Science Meets Ethnic Diversity: Ian McGonigle Interviews GenomeAsia100k Scientific Chairman Stephan Schuster. *Genetics Research* 101(e5), 1–6. doi:10.1017/S001667231800006X.

McGonigle, Ian V. 2015. "Jewish Genetics" and the "Nature" of Israeli Citizenship. *Transversal: Journal for Jewish Studies* 13(2), 90–102.

McGonigle, Ian V. 2017. Spirits and Molecules: Ethnopharmacology and Symmetrical Epistemological Pluralism. *Ethnos: Journal of Anthropology* 82(1), 139–164.

McGonigle, Ian V., and Ruha Benjamin. 2016. The Molecularization of Identity: Science and Subjectivity in the 21st Century. *Genetics Research* 98(e12). https://doi.org/10.1017/S0016672316000094.

McGonigle, Ian V., and Lauren W. Herman. 2015. Genetic Citizenship: DNA Testing and the Israeli Law of Return. *Journal of Law and the Biosciences* 2(2), 469–478.

McGonigle, Ian V., and Noam Shomron. 2016. Privacy, Anonymity, and Subjectivity, in Genomic Research. *Genetics Research* 98(e2). https://doi.org/10.1017/S0016672315000221.

Mol, Annemarie. 1999. Ontological Politics: A Word and Some Questions. *Sociological Review* 47(S1), 74–89.

Mor, Eytan, and Hagai Boas. 2005. Organ Trafficking: Scope and Ethical Dilemma. *Current Diabetes Reports* 5(4), 294–299.

Morris-Reich, Amos. 2006. Arthur Ruppin's Concept of Race. *Israel Studies* 11(3), 1–30.

Mosko, Mark S. 2015. Unbecoming Individuals: The Partible Character of the Christian Person. *HAU: Journal of Ethnographic Theory* 5(1), 361–393.

Mozersky, Jessica, and Galen Joseph. 2010. Case Studies in the Co-production of Populations and Genetics: The Making of "At Risk Populations" in BRCA Genetics. *BioSocieties* 5(4), 415–439.

Nachshoni, Kobi. 2014. Chief Rabbi: Stop Allowing Non-Jews to Make Aliyah. *YNet News*, November 3. http://www.ynetnews.com/articles/0,7340,L-4587242,00.html.

Nassar, Adid. 2019. In His Machiavellian Designs, Bassil Is Using Racist Discourse. *Arab Weekly*, June 16. https://thearabweekly.com/his-machiavellian-designs-bassil-using-racist-discourse.

National Academy of Sciences (NAS). 2011. *Toward Precision Medicine: Building a Knowledge Network for Biomedical Research and a New Taxonomy of Disease*. Washington, DC: NAS.

National Transplant Center. 2018. The National Transplant Center—One Center for All. Accessed February 26, 2018. https://www.adi.gov.il/en/about-us/.

Navon, Daniel, and Gil Eyal. 2016. Looping Genomes: Diagnostic Change and the Genetic Makeup of the Autism Population. *American Journal of Sociology* 121(5), 1416–1471.

Nesis, Lawrence S. 1970. Who Is a Jew? Shalit v. Minister of Interior et al. The Law of Return (Amendment No. 2). *Manitoba Law Journal* 4, 53–59.

NLGIP (National Laboratory for the Genetics of Israeli Populations). 2019a. Home. Accessed November 30, 2019. http://www.tau.ac.il/medicine/NLGIP.

NLGIP (National Laboratory for the Genetics of Israeli Populations). 2019b. Catalog. Accessed November 30, 2019. http://www.tau.ac.il/medicine/NLGIP/catalog.htm.

NLGIP (National Laboratory for the Genetics of Israeli Populations). 2019c. Contribution. Accessed November 30, 2019. http://www.tau.ac.il/medicine/NLGIP/contrib.htm.

NLGIP (National Laboratory for the Genetics of Israeli Populations). 2019d. Policy. Accessed November 30, 2019. http://www.tau.ac.il/medicine/NLGIP/policy.htm.

Ong, Aihwa. 1999. *Flexible Citizenship: The Cultural Logics of Transnationality*. Durham, NC: Duke University Press.

Ostrer, Harry. 2001. A Genetic Profile of Contemporary Jewish Populations. *Nature Reviews Genetics* 2(11), 891–898.

Ostrer, Harry, and Karl Skorecki. 2013. The Population Genetics of the Jewish People. *Human Genetics* 132(2), 119–127.

Palmié, Stephan. 2007. Genomics, Divination, Racecraft. *American Ethnologist* 34(2), 205–222.

Perez, Bien. 2017. China's Precision Medicine Initiative Gets Lift from Latest Genomics Company Funding. *South China Morning Post*, May 2. https://www.scmp.com/tech/china-tech/article/2092362/chinas-precision-medicine-initiative-gets-lift-latest-genomics.

Petryna, Adriana. 2013. *Life Exposed: Biological Citizens after Chernobyl*. Princeton, NJ: Princeton University Press.

Prainsack, Barbara. 2006. "Negotiating Life": The Regulation of Human Cloning and Embryonic Stem Cell Research in Israel. *Social Studies of Science* 36(2), 173–205.

Prainsack, Barbara. 2007. Research Populations: Biobanks in Israel. *New Genetics and Society* 26(1), 85–103.

Prainsack, Barbara. 2014a. Understanding Participation: The "Citizen Science" of Genetics. In *Genetics as Social Practice*, edited by Barbara Prainsack, Gabriele Werner-Felmayer, and Silke Schicktanz, 147–164. Farnham, UK: Ashgate.

Prainsack, Barbara. 2014b. Personhood and Solidarity: What Kind of Personalized Medicine Do We Want? *Personalized Medicine* 11(7), 651–657.

Prainsack, Barbara, and Yael Hashiloni-Dolev. 2009. Religion and Nationhood. In *Handbook of Genetics and Society*, edited by Paul Atkinson, Peter Glasner, and Margaret Lock, 404–421. London: Routledge.

Precision Medicine World Conference. 2018. All of Us Research Program Aims to Create Diverse Research Data Resource. *Precision Medicine World Conference*. Accessed July 30, 2018. https://www.pmwcintl.com/all-of-us-research-program-aims-to-create-diverse-research-data-resource/.

Qatar Biobank. 2015a. What Is Qatar Biobank? Accessed October 25, 2015. http://www.qatarbiobank.org.qa/about-us/what-is-qatar-biobank.

Qatar Biobank. 2015b. Research Focus Areas. Accessed October 25, 2015. http://www.qatarbiobank.org.qa/research/research-focus-area.

Qatar Biobank. 2015c. Pilot Phase Findings. Accessed November 28, 2015. http://www.qatarbiobank.org.qa/research/pilot-phase-findings.

Qatar Biobank. 2015d. Key Figures. Accessed November 28, 2015. http://www.qatarbiobank.org.qa/research/key-figures.

Qatar Biobank. 2015e. Research. Accessed November 28, 2015. http://www.qatarbiobank.org.qa/research.

Qatar Biobank. 2018. Qatar Biobank Reaches 15,000 Participants Milestone. Accessed September 29, 2019. https://www.qatarbiobank.org.qa/app/media/1858.

Qatar Biobank. n.d.-a. General Information Leaflet. Accessed November 28, 2015. http://dlnkk4xtshu10a.cloudfront.net/app/media/436.

Qatar Biobank. n.d.-b. Pilot Phase Report Summary. Accessed November 28, 2015. http://dlnkk4xtshu10a.cloudfront.net/app/media/1288.

Qatar Biobank. n.d.-c. Biobank Report. Accessed November 28, 2015. http://dlnkk4xtshu10a.cloudfront.net/app/media/1301.

Qatar Biobank. n.d.-d. Islamic View for Participation in the Qatar Biobank. Accessed November 28, 2015. http://dlnkk4xtshu10a.cloudfront.net/app/media/1165.

Qatar Foundation. 2014. Mapping the Qatari Genome Points Way to Prevention of Inherited Diseases. January 18. https://www.qf.org.qa/news/mapping-the-qatari-genome-points-way-to-prevention-of-inherited-diseases.

Qatar Foundation. 2016. Mapping the Roots of Qatar. *Monthly Magazine of Qatar Foundation* (94). https://originsofdoha.files.wordpress.com/2017/03/mapping-the-roots-of-qatar-qf-article.pdf.

Qatar Genome. 2018. Qatar Genome and Qatar Biobank Lead Efforts to Produce the First Qatari Gene Chip. Accessed September 29, 2019. https://qatargenome.org.qa/node/138.

Rabinowitz, Aaron. 2019. Israel's Rabbinical Courts Begin to Recognize DNA Tests, Potentially Opening Gateway to Proving Jewishness. *Ha'Aretz*, September 1. https://www.haaretz.com/israel-news/.premium-will-dna-testing-become-the-gateway-to-proving-jewishness-1.7772764?=&ts=_1579942049047.

Rabinowitz, Aaron. 2020. Israeli High Court Allows DNA Testing to Prove Judaism. *Ha'Aretz*, January 24. https://www.haaretz.com/israel-news/.premium-israeli-high-court-allows-dna-testing-to-prove-judaism-1.8439615.

Rager, Netta. 2015. Thesis Proposal: Protecting Privacy in Personalized Genomic Information. Noam Shomron Lab, Sackler School of Medicine, Tel Aviv University.

Rajan, Kaushik Sunder. 2006. *Biocapital: The Constitution of Post-Genomic Life*. Durham, NC: Duke University Press.

Ram, Uri. 2008. *The Globalization of Israel: McWorld in Tel Aviv, Jihad in Jerusalem*. New York: Routledge.

Reardon, Jenny. 2004. *Race to the Finish: Identity and Governance in an Age of Genomics*. Princeton, NJ: Princeton University Press.

Reardon, Jenny. 2011. The Democratic, Anti-Racist Genome? Technoscience at the Limits of Liberalism. *Science as Culture* 21(1), 25–47.

Reardon, Sara. 2015. US Tailored-Medicine Project Aims for Ethnic Balance. *Nature* 523(7561), 391–392.

Regalado, Antonio. 2018. 2017 Was the Year Consumer DNA Testing Blew Up. *MIT Technology Review*, February 12. https://www.technologyreview.com/s/610233/2017-was-the-year-consumer-dna-testing-blew-up/.

Richmond, Nancy C. 1993. Israel's Law of Return: Analysis of Its Evolution and Present Application. *Penn State International Law Review* 12(1), Article 4.

Rodriguez-Flores, Juan L., Khalid Fakhro, Neil R. Hackett, Jacqueline Salit, Jennifer Fuller, Francisco Agosto-Perez, Maey Gharbiah, et al. 2014. Exome Sequencing Identifies Potential Risk Variants for Mendelian Disorders at High Prevalence in Qatar. *Human Mutation* 35(1), 105–116.

Rose, Nikolas. 2007. *The Politics of Life Itself: Biomedicine, Power, and Subjectivity in the Twenty-First Century*. Princeton, NJ: Princeton University Press.

Rosenberg, Noah A., Eilon Woolf, Jonathan K. Pritchard, Tamar Schaap, Dov Gefel, Issac Shpirer, Uri Lavi, et al. 2001. Distinctive Genetic Signatures in the Libyan Jews. *Proceedings of the National Academy of Sciences* 98(3), 858–863.

Rubinstein, Amnon. 2016. The Lie Behind "Genetic Citizenship." *Israel Hayom*, June 24.

Salibi, Kamal. 1988. *A House of Many Mansions: The History of Lebanon Reconsidered*. Berkeley: University of California Press.

Sand, Shlomo. 2009. *The Invention of the Jewish People*. Brooklyn, NY: Verso.

Sapir, Gidon. 2006. How Should a Court Deal with a Primary Question That the Legislature Seeks to Avoid? The Israeli Controversy over Who Is a Jew as an Illustration. *Vanderbilt Journal of Transnational Law* 39(4), 1233–1239.

Sarna, Jonathan. 2011. Ethnicity and Beyond. In *Ethnicity and Beyond: Theories and Dilemmas of Jewish Group Demarcation*, edited by E. Lederhendler, 108–114. Oxford: Oxford University Press.

Schwartz-Marín, Ernesto, and Arely Cruz-Santiago. 2016. Forensic Civism: Articulating Science, DNA and Kinship in Contemporary Mexico and Colombia. *Human Remains and Violence: An Interdisciplinary Journal* 2(1), 58–74.

Schwartz-Marín, Ernesto, and Eduardo Restrepo. 2013. Biocoloniality, Governance, and the Protection of "Genetic Identities" in Mexico and Colombia. *Sociology* 47(5), 993–1010.

Schwartz-Marín, Ernesto, and Irma Silva-Zolezzi. 2010. The Map of the Mexican's Genome: Overlapping National Identity, and Population Genomics. *Identity in the Information Society* 3(3), 489–514.

Seeman, Don. 2010. *One People, One Blood: Ethiopian-Israelis and the Return to Judaism*. New Brunswick, NJ: Rutgers University Press.

Senor, Dan, and Saul Singer. 2009. *Start-Up Nation: The Story of Israel's Economic Miracle*. New York: Twelve.

Shafir, Gershon, and Yoav Peled. 2002. *Being Israeli: The Dynamics of Multiple Citizenship*. Cambridge: Cambridge University Press.

Shamir, Ronen. 2013. *Current Flow: The Electrification of Palestine*. Palo Alto, CA: Stanford University Press.

Shapin, Steven. 2000. Trust Me. *London Review of Books* 22(9), 15–17. https://www.lrb.co.uk/the-paper/v22/n09/steven-shapin/trust-me.

Shen, Peidong, Tal Lavi, Toomas Kivisild, Vivian Chou, Deniz Sengun, Dov Gefel, Issac Shpirer, et al. 2004. Reconstruction of Patrilineages and Matrilineages of Samaritans and Other Israeli Populations from Y-Chromosome and Mitochondrial DNA Sequence Variation. *Human Mutation* 24(3), 248–260.

Shomron, Noam. 2013. *Deep Sequencing and Data Analysis*. Edited by Noam Shomron. New York: Humana Press.

Shomron, Noam. 2014. Prioritizing Personalized Medicine. *Genetics Research* 96(e7).

Shomron, Noam. 2017a. Should You Read Your DNA? TEDxTelAvivUniversity. Accessed January 10, 2017. https://youtu.be/NgVwPj54TEo.

Shomron, Noam. 2017b. Big Data and Genomics: Halting the Spread of Breast Cancer. Nano World Cancer Day at Tel Aviv University. Accessed February 3, 2017. https://www.youtube.com/watch?v=yGf5JTwvsIQ.

Sidra. 2014. Sidra 5-Year Strategic Plan. Accessed November 28, 2015. http://www.sidra.org/wp-content/uploads/2014/12/sidras-five-year-strategic.pdf.

Sidra. 2015a. Poster from Functional Genomics Symposium. Accessed November 28, 2015. http://events.sidra.org/event/functional-genomics-symposium/.

Sidra. 2015b. Sidra Fact Sheet. Accessed November 28, 2015. http://www.sidra.org/fact-sheet-patient-centric-technology/.

Sidra. 2016. Functional Genomics Symposium. Accessed December 18, 2016. http://events.sidra.org/event/functional-genomics-symposium-2016/.

Siegal, Gil. 2015. Genomic Databases and Biobanks in Israel. *Journal of Law, Medicine & Ethics* 43(4), 766–775.

Silverstein, Richard. 2013. Birthright, Israeli Government Demand DNA Tests to Prove Jewishness. *Tikkun Olan Blog*, August 3. http://www.richardsilverstein.com/2013/08/04/birthright-israeli-government-demand-dna-tests-to-prove-jewishness/.

Skorecki, Karl, Sara Selig, Shraga Blazer, Robert Bradman, Neil Bradman, P. J. Waburton, Monica Ismajlowicz, et al. 1997. Y Chromosomes of Jewish Priests. *Nature* 385(6611), 32.

Strathern, Marilyn. 1988. *The Gender of the Gift: Problems with Women and Problems with Society in Melanesia*. Berkeley: University of California Press.

Sun, Shirley. 2017. *Socio-economics of Personalized Medicine in Asia*. London: Routledge.

TallBear, Kim. 2013a. *Native American DNA: Tribal Belonging and the False Promise of Genetic Science*. Minneapolis: University of Minnesota Press.

TallBear, Kim. 2013b. Genomic Articulations of Indigeneity. *Social Studies of Science* 43(4), 509–534.

Tel Aviv University. 2013. TAU Research Finds Gene That Potentially Predicts Antidepressant Response. December 11. https://english.tau.ac.il/news/gene_antidepressant.

Thomas, Mark G., Tudor Parfitt, Deborah A. Weiss, Karl Skorecki, James F. Wilson, Magdel le Roux, Neil Bradman, and David B. Goldstein. 2000. Y Chromosomes Traveling South: The Cohen Modal Haplotype and the Origins of the Lemba—the "Black Jews of Southern Africa." *American Journal of Human Genetics* 66(2), 674–686.

Thomas, Mark G., Michael E. Weale, Abigail L. Jones, Martin Richards, Alice Smith, Nicola Redhead, Antonio Torroni, et al. 2002. Founding Mothers of Jewish Communities: Geographically Separated Jewish Groups Were Independently Founded by Very Few Female Ancestors. *American Journal of Human Genetics* 70(6), 1411–1420.

1000 Genomes Project. 2015. A Deep Catalog of Human Genetic Variation. Accessed November 16, 2015. http://www.1000genomes.org/about.

Troen, Ilan, and Carol Troen. 2019. Indigeneity. *Israel Studies* 24(2), 17–32.

Tupasela, Aaro, and Sakari Tamminen. 2015. Authentic, Original, and Valuable: Stabilizing the Genetic Identity in Non-human and Human Populations in Finland. *Studies in Ethnicity and Nationalism* 15(3), 411–431.

Viveiros de Castro, Eduardo. 2013. The Relative Native. *HAU: Journal of Ethnographic Theory* 3(3), 473–502.

Vogel, Carol. 2013. Art, from Conception to Birth in Qatar: Damien Hirst's Anatomical Sculptures Have Their Debut. *New York Times*, October 7. http://www.nytimes.com/2013/10/08/arts/design/damien-hirsts-anatomical-sculptures-have-their-debut.html.

Wagner, Roy. 1991. The Fractal Person. In *Big Men and Great Men: Personifications of Power in Melanesia*, edited by Maurice Godelier and Marilyn Strathern, 159–173. Cambridge: Cambridge University Press.

Weill Cornell Medicine–Qatar. 2009. Arabian Oryx Genome Data. Accessed April 28, 2018. https://qatar-weill.cornell.edu/research/research-highlights/arabian-oryx-genome-sequence/arabian-oryx-genome-data.

Weill Cornell Medicine–Qatar. 2013. Qatar Genome Project (QGP). Accessed November 20, 2015. https://qatar-weill.cornell.edu/media/reports/2013/qatariGenome.html.

Weingrod, Alex. 2015. Afterword. In *Toward an Anthropology of Nation Building and Unbuilding in Israel*, edited by Fran Markovitz, Stephen Sharot, and Moshe Shokeid, 317–322. Lincoln: University of Nebraska Press.

Weiss, Meira. 2004. *The Chosen Body: The Politics of the Body in Israeli Society*. Palo Alto, CA: Stanford University Press.

Wheelwright, Jeff. 2013. Defining Jews, Defining a Nation: Can Genetics Save Israel? *The Atlantic*, March 14. http://www.theatlantic.com/global/archive/2012/03/defining-jews-defining-a-nation-can-geneticssave-israel/254428/.

White House. Office of the Press Secretary. 2015. Factsheet: President Obama's Precision Medicine Initiative. January 30. https://www.whitehouse.gov/the-press-office/2015/01/30/fact-sheet-president-obama-s-precision-medicine-initiative.

Zeiger, Asher. 2013. Russian-Speakers Who Want to Make *Aliya* Could Need DNA Test. *Times of Israel*, July 29. https://www.timesofisrael.com/russian-speakers-who-want-to-immigrate-could-need-dna-test/.

Zerubavel, Yael. 1995. *Recovered Roots: Collective Memory and the Making of Israeli National Tradition*. Chicago: University of Chicago Press.

Žižek, Slavoj. 1993. *Tarrying with the Negative*. Durham, NC: Duke University Press.

INDEX

Note: Page numbers in *italics* indicate figures and tables.

Printed in the United States
by Baker & Taylor Publisher Services